Lecture Notes in Earth Sciences 82

Springer-Verlag Berlin Heidelberg GmbH

Tore M. Løseth

Submarine Massflow Sedimentation

Computer Modelling and Basin-Fill Stratigraphy

With 29 Figures and 3 Tables

 Springer

Author

Tore M. Løseth
University of Bergen
Geological Institute
N-5007 Bergen, Norway
E-mail: Tore.Loseth@geol.uib.no

"For all Lecture Notes in Earth Sciences published till now please see final pages of the book"

Cataloging-in-Publication data applied for

Die Deutsche Bibliothek - CIP-Einheitsaufnahme

Løseth, Tore M.:
Submarine massflow sedimentation : computer modelling and basin fill stratigraphy / Tore M. Løseth. - Berlin ; Heidelberg ; New York ; Barcelona ; Budapest ; Hong Kong ; London ; Milan ; Paris ; Singapore ; Tokyo : Springer, 1999
(Lecture notes in earth sciences ; 82)

ISSN 0930-0317
ISBN 978-3-540-65057-7   ISBN 978-3-540-49700-4 (eBook)
DOI 10.1007/978-3-540-49700-4

© Springer-Verlag Berlin Heidelberg 1999
Originally published by Springer-Verlag Berlin Heidelberg New York in 1999.

The use of general descriptive names, registered names, trademarks, etc. in this publication does not imply, even in the absence of a specific statement, that such names are exempt from the relevant protective laws and regulations and therefore free for general use.

Typesetting: Camera ready by author
SPIN: 10692029   32/3142-543210 - Printed on acid-free paper

Preface

Thick successions of deep-marine massflow deposits are common in the geological record and many of them are important petroleum reservoirs, notably in the continental shelf area of North-Western Europe. The topic of the present volume is a computerized modelling of siliciclastic sediment gravity-flow sedimentation in a deepwater basin in the stratigraphic framework of two-dimensional basin-fill architecture. Stratigraphic modelling is an important key to an understanding of the large-scale internal variation, depositional architecture and geometry of a sedimentary system, which are crucial physical aspects in the analysis of modern systems and the subsurface exploration of the ancient ones. Computer simulations have also become an important tool of the geological sequence-stratigraphic analysis, whose methodological aim is to decipher the principal dynamic aspects of a sedimentary basin's history from the large-scale depositional trends and architectural patterns of the basin-fill succession.

Despite its great potential importance, the stratigraphic modelling of submarine depositional systems has largely been still-born, until recently, because of the lack of an adequate numerical concept. The present approach to the computer modelling of a massflow-dominated system, though necessarily a physical simplifaction, is fully original and has many important advantages in comparison to the previous numerical attempts. The algorithm used is based on a process-oriented, dynamic slope concept, which has no comparable advanced parallel in the existing computer programs.

The contents of the book are organized into eight chapters. Chapter 1 is a short introduction to stratigraphic simulation models, focusing on the DEMOSTRAT program upon which the present approach has been founded. (The latter program, now widely available, is an assembly of forward simulation algorithms based on the physical law of diffusion.) Chapter 2 addresses the problem of submarine slope instability. Chapter 3 gives a comprehensive review of the triggering factors and mechanics of submarine sediment-gravity flows (debrisflows and turbidity currents), and discusses the main types of depositional systems dominated by such massflows. Chapter 4 is a review of the existing numerical concepts of stratigraphic modelling, highlighting the difference between geometrical and process-based models of a depositional system. Chapter 5 describes the algorithm developed by the present author, with an emphasis on the formulation of the component algorithms for slope instability and massflow sedimentation. Chapter 6 reviews the results of example computer simulations and discusses briefly their basinal implications. The examples demonstrate their great potential usefulness of the program and show as to how the program's algorithms function within the DEMOSTRAT software. Chapter 7 is a discussion of the present algorithm's advantages, shortcomings and possible improvements. Chapter 8 closes the volume with a summary of the main conclusions and practical implications. The volume has a small appendix on the numerical procedure used and includes also a comprehensive list of literature references that will certainly be useful to many readers.

This is a book for geoscientists whose research is concerned with sequence stratigraphy and basin analysis, which are topics now occupying many desks in the academic institutions and petroleum companies alike. The language of the book is relatively simple, essentially non-mathematical, and the richly illustrated text should thus be fully comprehensive to any researcher with an elementary knowledge of mathematics and mechanics.

The author is indebted to Dr. Jan C. Rivenæs (Norsk Hydro Research Centre, Bergen) for introducing me to stratigraphic modelling, for many useful recommendations, for implementing the present algorithms into the DEMOSTRAT framework and for help with the document preparation system. The author also wants to thank Professor Wojtek Nemec (University of Bergen) for improving the manuscript and correcting the language and Dr. William Helland-Hansen (University of Bergen) for many profitable discussions. Norsk Hydro is acknowledged for financial support.

Contents

Chapter 1

Introduction

1.1   The Aim and Methods

The general aim of the present study was to develop algorithms for massflow sedimentation within computer stratigraphic simulations of large-scale 2-D basin-fill architecture, and to fit them into the numerical framework of DEMOSTRAT software. The study was carried out at the Geological Institute, University of Bergen, in 1996-98, with part of the work done at the Norsk Hydro Research Centre in Bergen.

In the computer modelling of dynamic stratigraphy, there are usually two approaches to model the erosion, transport, and deposition of sediments: *geometric modelling* or *process-based modelling* (see below, section 1.2). The process-based models include those based on slope dynamics and those based on fluid-flow dynamics (section 1.2). The work that has up to date been done to simulate sediment gravity flow sedimentation, in terms of both geometric and dynamic-slope models, is relatively simple (see review in Chapter 4). In most of the models, the triggering, movement and deposition of massflows are defined by *a priori* defined angles. This means that even the dynamic-slope models employ a geometrical basis to model sediment-gravity flows. Many of the models based on fluid-flow dynamics are relatively good, but both the time scale considered and the detailed input reqired render them little useful for stratigraphic modelling purposes (see section 1.2 and Chapter 4). In short, there is no advanced computer model for a dynamic stratigraphic simulation of sediment gravity-flow sedimentation.

The aim of this study was to develop such a model. This has been done by using the existing physical principles of massflow triggering and deposition mechanics. The source erosion and the transport and deposition of sediment by massflows are considered in terms of the law of diffusion, because the dynamic-slope model employed in DE-MOSTRAT is based on diffusion equations. The DEMOSTRAT program is explained in some detail in section 1.4. The present elaboration of DEMOSTRAT renders this program capable of modelling the 2-D longitudinal geometry of sediment bodies resulting from prolonged massflow deposition, including the effects of sediment bypass.

DEMOSTRAT is a stratigraphic computer model based on dynamic-slope concepts (section 1.2), which means that the stratigraphic resolution in this model is approximately the same as in seismic stratigraphy. The program models siliciclastic basin-fill architecture in terms of a 2-D longitudinal cross-section (average for the depositional slope's strike), using discrete time increments (typically of the order of 50,000-100,000 yrs) and discrete pints in the 2-D space. The model can be termed a sediment-budget model, because the available sediment (either externally supplied or derived by intrabasinal erosion) is being distributed within the same 2-D basinal section. Despite its limitations, the model is thought to provide a better approach to the stratigraphic modelling of massflow sedimentation than those offered by the existing other programs.

1.2   Stratigraphic Simulation Models

In natural sciences, a model is a schematic or simplified description that attempts to portray natural processes. To simulate is to perform experiments on such a model (Harbaugh & Bonham-Carter, 1977). An attempt to model natural phenomena quantitatively requires assumptions, idealized conditions and necessary simplifications; a defensible situation if the user of the model is aware of these and of the limitations they impose on the results (Cant, 1991). A consequence is that the better the researcher understands the limitations of a model, the more reliable are the conclusions based on the results of

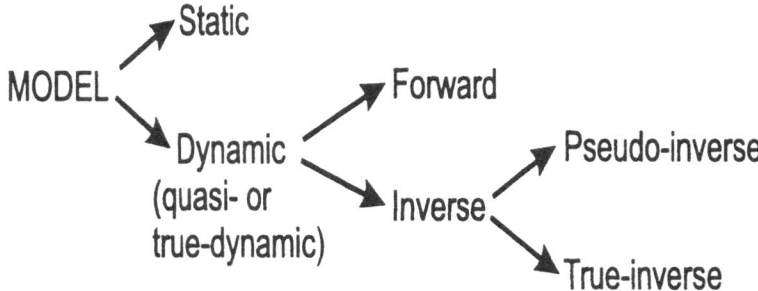

Figure 1.1: Simulation models classified according to the time parameter.

the model.

Simulation models may be either dynamic or static (Fig. 1.1). In static models, time is not a physical variable. In dynamic models, time is a variable parameter, and the behaviour of a dynamic system is reproduced as it evolves through time (Bonham-Carter & Harbaugh, 1968). Dynamic models can further be divided in quasi-dynamic and true-dynamic models (Rivenæs, 1993). Quasi-dynamic models are "forward" models, but the time itself is not a variable that affects the results (e.g., when porosity-curves are used to model compaction, the same result is obtained no matter how long the compaction has taken place). In true dynamic models, time is an important factor. Many forward models are hybrids, consisting of elements of both quasi- and true-dynamic concepts.

There are two categories of dynamic simulation approach, namely forward modelling and inverse modelling (Fig. 1.1). In a forward modelling, one makes predictions (responses) based on causative agents (processes), and such models are often termed process-response models (Whitten, 1964). They attempt to model the interacting time-dependent processes that operate within a system, following in the inferred footsteps of "Mother Nature", and are used in a predictive manner. Inverse models include two categories: pseudo-inverse models and true inverse models (cf. Lerche, 1990). Pseudo-inverse models use forward modelling to find the possible values of variables that would lead to a certain particular (known) result (Bice, 1988). This ultimately requires a large number of iterative simulations in which the forward model is progressively adjusted, such that the predictions would eventually match the observations in a satisfactory manner. In contrast, a true inverse model uses established measurements to reveal the dynamic evolution of the process considered. A dynamic model involves one dimension in time and up to three dimensions in space.

Stratigraphic simulation models are forward dynamic models that focus on the processes that progressively fill a basin with sediment, including the spatial distribution and interrelationships of sediment sequences and facies in a time-stratigraphic framework (Rivenæs, 1993). All other relevant processes (e.g., eustasy and tectonic activity) are either included directly in the model or represented as its initial or boundary conditions.

On the basis of the erosion-deposition algorithm employed, stratigraphic simulation models can be divided into geometric and process-based models (see for instance Rivenæs, 1992). **Geometrical models** are based on empirical observations on deposits geometries. Hence, it is the depositional geometries that are modelled as responses, without dealing with the actual processes that created these particular geometries. In contrast, **process-based models** use physical assumptions about the sedimentation process to model the erosion, transport and deposition of the sediment mass. The advantage of process-based models is that they do not specify the resulting depositional geometries in advance, but use algorithms that attempt to imitate sediment erosion, transport and deposition. The disadvantage is that these processes are physically complex and not always fully understood, hence some simplifying assumptions are invariably required in order to represent the sedimentation process mathematically.

Process-based models can further be divided into fluid-flow models and dynamic-slope models (Rivenæs, 1992), dependent upon the actual temporal and spatial scales used. **Fluid-flow** models simulate the sedimentation process at a relatively detailed temporal and spatial level (for instance, the deposition of individual channel-fills, individual turbidites or individual mouth bars), whereas **dynamic-slope** models consider the sedimentation process at a larger-scale level, corresponding to the basin's dimensions, but do not resolve small-scale topographic irregularities, individual facies elements and short-lived events (Rivenæs, 1993).

4

A different approach to the simulation of basin-fill architecture is used in stochastic modelling (see reviews by Dubrule, 1988; Haldorsen & Lake, 1990). Such algorithms model only the possible spatial distribution, not temporal evolution, and are thus static simulations. The lack of dynamic component in these models renders them rather unsuitable for understanding stratigraphic relationships (Rivenæs, 1993).

1.3 Applications

Stratigraphic computer-simulation models are useful for four main reasons (Waltham, 1992). Firstly, they are excellent learning aids, when used interactivley, allowing the user to gain an insight into the ways in which the sedimentation process alone, or in combination with tectonic and/or climatic factors, creates the basin-fill stratigraphy and architecture. Secondly, numerical models can be used to estimate unknown parameters. For example, the geometrics seen in seismic sections may be a result of the interaction of eustatic sea-level changes, rates of sedimentation, and the rates and direction of tectonic movement. Simulations can vary these parameters until a favourable match between the simulated section and the interpreted seismic cross-section is reached. On this basis, the most likely ranges of the sediment accumulation rate, subsidence rate and eustatic sea-level changes responsible for the observed geometries can be estimated (Helland-Hansen *et al.*, 1988). Thirdly, numerical stratigraphic modelling is also used to interpolate between outcrops, or well sites, by finding a model cross-section that closely matches the local data and assuming that the model applies as well to the unknown portions of the basinal space. This gives the possibility to predict the architecture and lithology of the basin-fill and to test rapidly various exploration scenarios in frontier basins or their parts. Finally, numerical simulation can be used to test conceptual geological models by making predictions that can directly be compared with the actual data. Examples of various stratigraphic simulations done with the use of DEMOSTRAT are given in Chapter 6

The quality of a stratigraphic simulation depends upon the reliability of the input parameters (as it is often said, crap in – crap

5

out) and on the capability of the computer program to account for the crucial aspects of the basin-fill process and its controls (Aigner *et al.*, 1990). Implicit in every application is the assumption that the mathematical model realistically describes the basin-fill processes, but considering the simplicity of most such models, they can be expected to approximate the real world only roughly. This, however, is not necessarily a reason for disqualification. Once a stratigraphical model has been carefully deviced, it is liable to further improvements, because its parts can readily be examined and then modified and expanded as our understanding of geological processes involved advances (Horowitz, 1976). Hence, even if a simulation study fails to yield a valid picture of the basin-fill succession, the results often provide the researcher with valuable new insight into the stratigraphic problems being investigated (Harbaugh, 1966). Another important point to be emphasized is that the quality of basinal predictions made on the basis of a simulation will depend also on the knowledge of the mathematical model; the results of "black-box" simulations may easily be misinterpreted or lead to misunderstandings.

1.4 DEMOSTRAT

The present study is a direct extension of the dynamic-slope, two-dimensional stratigraphic simulation model named DEMOSTRAT (DEpositional MOdelling of STRATigraphy), developed in 1990 - 1993 by Rivenæs (1993). The model is a forward computer program that contains elements of both quasi-dynamic and true dynamic models. The DEMOSTRAT program is briefly introduced in this section, as the background for the present study.

1.4.1 Numerical Framework

The numerical framework of the simulator is a set of discrete points (node points), in both time and 2-D space (dip section, perpendicular to the basin margin). A schematic illustration of this is shown in Figure 1.2. The simulation in time is divided into equal-time steps, and in space into equal-width columns (Fig. 1.2 A & B).

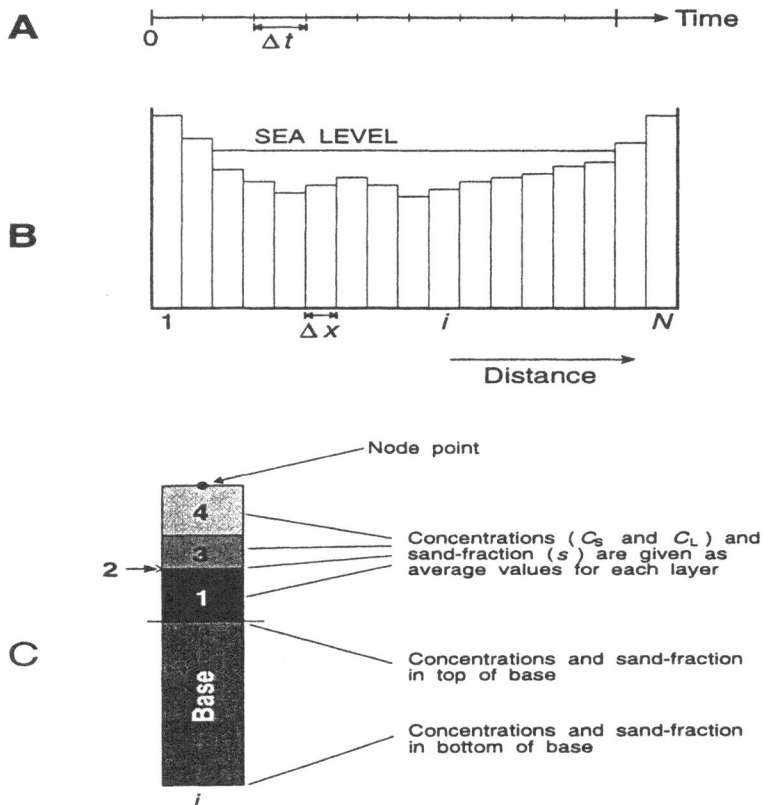

Figure 1.2: The numerical framework of DEMOSTRAT simulation (from Rivenæs, 1993). (A) The time span considered is divided into "steps", with equal time-lengths, Δt. (B) The basin consists of equal-width, Δx spaced columns (1 ... N) with the centre of each column being a numerical node-point; the sea level is an important datum in the simulations. (C) Each column consists of a sedimentary "base" (substratum) overlain by successive sediment layers. The layers deposited are characterized by their thicknesses (which can be zero in case of erosion/nondeposition, as illustrated by layer 2 here), their sand-fraction s, and their grain concentrations C_S (sand) and C_L (mud). The base has its own analogous characteristics. Because the base can have great thickness, the program calculates both the sand-fraction (s) and the grain concentrations (C_S and/or C_L) at its bottom and top, with the intermediate values found by linear interpolation.

1.4.2   The Time-Step Loop

The form of the time-step loop employed is summarized in Figure 1.3. Below, it is described as to how each of the geological processes, or variables, is actually modelled.

Tectonic Movements

The casual processes of tectonic movements are very complicated, generally too complex to be represented in a numerical algorithm. In DEMOSTRAT, tectonic movements are set *a priori* in the input file, with only vertical movements possible.

Eustasy

Eustasy, or eustatic sea-level changes, are defined as the changes in the elevation of sea level on a world-wide basis, relative to a stationary datum (e.g., the centre of the Earth) (Burton *et al.*, 1987). Changes in the water volume and changes in the ocean-basin volume are the two possible causes of eustatic sea-level variations. These changes are due to such factors as the trapping of water in glaciers, lakes and magmas; an increase in the volume of ocean water owing to the generation of juvenile water; changes in the mean oceanic temperature; variations in the atmospheric moisture; desiccation of marginal epicontinental seas; variation in sediment influx; changes in mid-oceanic ridges, and possibly the Earth expansion and geoidal changes (Donovan & Jones, 1979). The eustasy function is pre-set *a priori* in DEMOSTRAT, and the user samples the sea-level curves in two steps: First a stepwise linear trend is traced through the curves inflection points, and then a sinuous trend is added. This method is similar to that described by Helland-Hansen *et al.* (1988), and is considered to provide the user with sufficient flexibility.

Erosion and Deposition

Equations of physical diffusion are used in DEMOSTRAT to model the process of sediment erosion, transport and deposition, which is an approach with a relatively long tradition in stratigraphic modelling

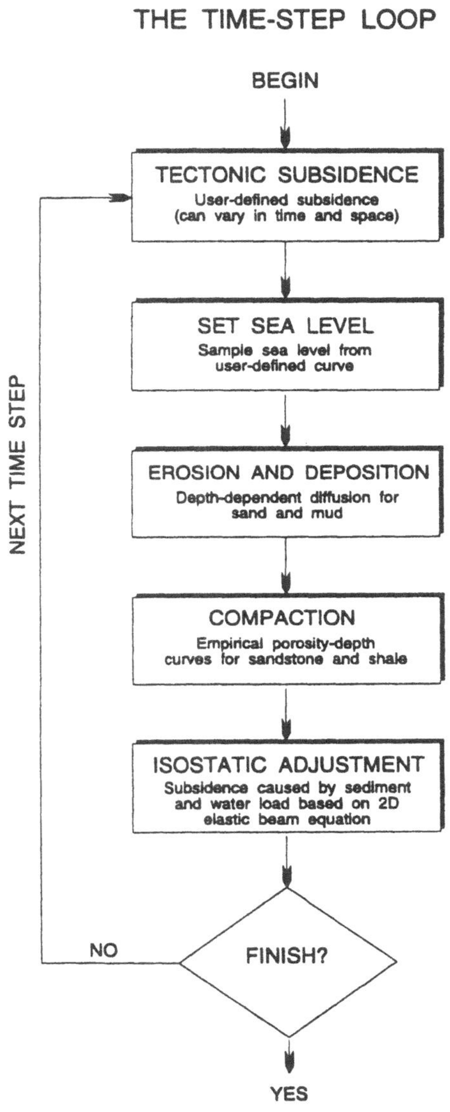

THE TIME-STEP LOOP

BEGIN

TECTONIC SUBSIDENCE
User-defined subsidence
(can vary in time and space)

SET SEA LEVEL
Sample sea level from
user-defined curve

EROSION AND DEPOSITION
Depth-dependent diffusion for
sand and mud

COMPACTION
Empirical porosity-depth
curves for sandstone and shale

ISOSTATIC ADJUSTMENT
Subsidence caused by sediment
and water load based on 2D
elastic beam equation

NEXT TIME STEP

NO

FINISH?

YES

Figure 1.3: The time-loop structure used in DEMOSTRAT (from Rivenæs, 1993). The processes are modelled sequentially within each time step. All modules, except for the erosion and deposition, can be bypassed if undesired.

9

(Rivenæs, 1992, see also section 4.2). The equation of a linear, homogeneous diffusion for one-type sediment of a particular topographic elevation is given as:

$$\frac{\partial h}{\partial t} = K \frac{\partial^2 h}{\partial x^2} \qquad (1.1)$$

where $h(x,t)$ is the elevation, t is time, x is the spatial in-plane dimension perpendicular to h, and K is the coefficient of diffusion, or transport. The equation states that the change in elevation with time is dependent upon the change of the topographic gradient. Further, erosion (i.e., a negative $\frac{\partial h}{\partial t}$) occurs when the change of gradient is negative (i.e., convex upward) and deposition (i.e., a positive $\frac{\partial h}{\partial t}$) occurs when the change of gradient is positive (i.e. concave upward). K can be made dependent upon the elevation (i.e. $K = k(h)$), and equation 1.1 then becomes

$$\frac{\partial h}{\partial t} = \frac{\partial}{\partial x}\left(K \frac{\partial h}{\partial x}\right) \qquad (1.2)$$

The erosion, transport and deposition are controlled by both the change in gradient and the elevation. The transport coefficient, K $[\frac{m^2}{year}]$, is a measure of the efficiency of sediment transport, indicating as to how easily the deposition and erosion can occur. In other words, they reflect the depositional capacity of the basinal environment. Now, the transport coefficients can be adjusted so that, to put it simply, erosion occurs above the base level and the deposition occures below the base level. The higher the K-values, the more easily will the sediment be transported. Thus, coefficient K is a very important quantity; alas, it is also very difficult to determine, particularly since each sediment type may have its own "mobility" characteristics. K is also a function of the scale considered, and apparently increases with the scale (cf. Anderson & Humphrey, 1990). DEMOSTRAT uses a dual-sediment equation which means that the distance in the basin to which the sediment is sheerly transported and where the deposition begins can be determined independantly for two sediment types (e.g., sand and mud) by using different K-curves for each of them.

In summary, a system of dual-sediment, depth-dependent diffusion equations has been implemented to be numerically solved and to perform erosion and deposition in DEMOSTRAT. This system of equations is a sophisticated and probably the strongest, most original aspect of the program. The basic diffusion equations for sand (1.3) and mud (1.4) in DEMOSTRAT are (for derivation, see Rivenæs, 1992):

$$AC_S\frac{\partial s}{\partial t} + C_S s\frac{\partial h}{\partial t} = \frac{\partial}{\partial x}\left(l_S K_S\frac{\partial h}{\partial x}\right) \tag{1.3}$$

$$-AC_L\frac{\partial s}{\partial t} + C_L(1-s)\frac{\partial h}{\partial t} = \frac{\partial}{\partial x}\left(l_L K_L\frac{\partial h}{\partial x}\right) \tag{1.4}$$

where A = the thickness of a "surface zone", C = the concentration of sediment grains (subscipts S for sand and L for mud), s = the bulk-volume fraction of sand, K = transport (diffusion) coefficient, l = sediment flux fraction in the transport coefficient, h = topographic height (elevation) measured from a fixed datum, t = time, and x = horizontal distance.

The values of A and K are selected by the user to solve the diffusion equations. The presence of the "surface layer", A, is necessary for the modelling of transport and sand/mud sorting, and its thickness cannot be zero (Rivenæs, 1992). The surface layer is for simplicity set to be constant in space and time.

Compaction

Compaction is the decrease of the thickness of a sediment body due to loading. Empirical porosity-depth relations are used in DEMOSTRAT to mimic the sedimentary response to the compaction process. It is possible to perform a simulation without compacting the sediments. In case of erosion, no decompaction is taken into account.

Porosity-depth-relationship is represented by a variety of empirical curves, which are also partially dependent upon the geographical location of the area. There is a general agreement that mud behaves differently than sand during compaction. Two possible curves for sand and three curves for mud are included in DEMOSTRAT (for details, see Rivenæs, 1993). If the simulations are performed without sediment compaction, the sediments are assumed to have initial

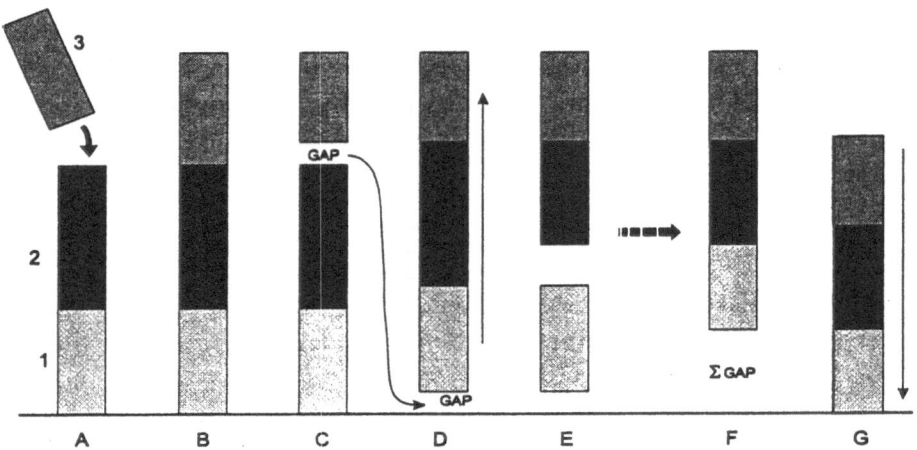

Figure 1.4: The concept of numerical compaction (from Rivenæs, 1993): A new layer (3) is added to the top of the sediment column (A). The layer is first compacted under its own weight. The depth of the top of this column (B) is known, so a new position of the bottom of the layer is computed (C), which gives a "gap" to the top of the underlying layer (here layer 2). This gap is transferred to the bottom of the column (D), with the layers temporarily "lifted". The calculation is repeated for each layer (E - F), and the resulting compacted column is then moved down to the datum surface (G).

porosities as in the Baldwin & Butler (1985) curves. The total porosity will depend upon the mud/sand ratio of sediment.

Compaction method is numerical and its principle is shown in Figure 1.4. In the layering model in DEMOSTRAT, each sediment layer is a mixture of mud and sand. In order to make the numerical compaction realistic, the algorithm separates the sediment into three hypothetical layers, with two layers of sand sandwiched by a layer of mud (see Figure 1.5). It is also possible to compact the "base", the substratum beneath the sediment layers deposited.

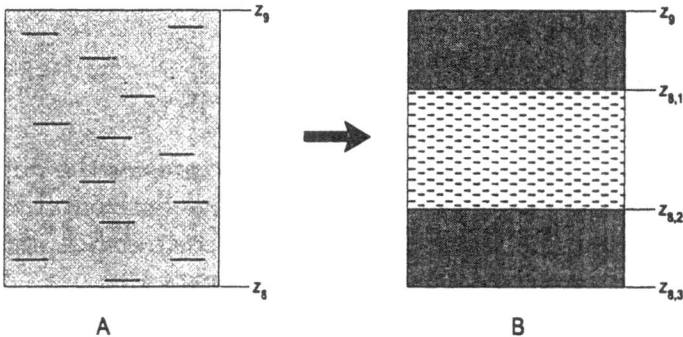

Figure 1.5: The sediment layer in DEMOSTRAT is considered to be a mixture of sand and mud (A). In the modelling of compaction, the layer is hypothetically divided into three layers, with one-half of the sand in the lower layer, all mud in the middle layer and the other half of the sand in the upper layer (B). The compaction is thus performed in three steps, for each layer separatively, which ensures a realistic amount of compaction ($z_{8.1}$; $z_{8.2}$; $z_{8.3}$) (from Rivenæs, 1993).

Isostasy

Since the lithosphere behaves elastically, it is able to bend when force systems or a load stress is applied to it. The subsidence due to loading is necessary to be incorporated in the model because of the influence it may exert on the sedimentation pattern. The concept of the flexuring of an elastic solid can be summarized briefly as follows:

1. Flexure may be a result of vertical forces, horizontal forces or a torque (bending) moment, possibly in combination. Horizontal loads are commonly neglected in geodynamical considerations.

2. The bending moment is the integration of the normal stresses over a cross-section of a crustal plate, acting over the plate's midline. The bending is related to the local radius of the plate's curvature by a coefficient called flexural rigidity. Flexural rigidity, D, is proportional to the cube of the elastic thickness, T, and further related to the Young modulus, E, and

13

Poisson ratio, p:

$$D = \frac{ET^3}{12(1 - p^2)} \tag{1.5}$$

A general equation of flexuring, which expresses the deflection of the platecrustal in terms of the vertical and horizontal loads, mechanical bending and flexural rigidity (Turcotte & Schubert, 1982), is employed in DEMOSTRAT and solved numerically.

1.5   Outline

The next two Chapters (2 and 3) discusses the general theory of sediment gravity flows, including the triggering mechanism, erosion, models of transport and style of deposition. This review forms the basis for the conceptual model that the author further formulates in mathematical terms. Chapter 4, is a review of the previous work on the issue of the numerical modelling of massflow processes. Chapter 5 explains the massflow model and its algorithms developed for this thesis. The results of some simulations are given in Chapter 6. Final discussions (Chapter 7) and conclusions (Chapter 8) close the book.

A table with the explanation of all symbols used in the volume is given in Appendix A. Appendix B describes the numerical method employed to calculate the integral used in the equation for a sediment's excess pore pressure (equation 2.14).

Chapter 2

Sediment Instability on Subaqueous Slopes

The most important triggering mechanism for sediment gravity flows, leading to the transport and deposition of siliciclastic sediment in deep-water environments, is sedimentary slope failure (e.g., Saxov & Nieuwenhuis, 1982; Brunsden & Prior, 1984; Colella & Prior, 1990; Maltman, 1994). The failure in sediment on a slope occurs when normal stress, or its downslope "shearing" component, in the sediment exceeds the resisting strength of the sediment (see section 2.1.1). Slope failure normally leads to a slide, which may further transform into a slump and/or turn into a sediment gravity flow (as discussed in Chapter 3). In this Chapter, the instability of sediment on subaqueous slopes is discussed.

2.1 Analysis of Sediment Instability on Subaqueous Slopes

Slope stability analysis is a well-known topic in the engineering discipline of soil mechanics. Soil engineers are interested in the mechanical problems of mass failure when it is necessary to stabilize natural slopes, or to create stable artificial ones, such as embarkments, road cuts and other excavations. In creating artificial slopes, which may be higher and/or steeper than the pre-existing natural slope, many "landslides" (mass movements) are accidentally produced by increasing critically the shear stresses in the slope's sediment mass (Carson & Kirkby, 1972). In order to avoid or to control these unwanted effects, slope stability analysis has been developed by engineers.

Below, two theories that can be combined to evaluate the stability of sediment on a slope are reviewed. The relevant boundary conditions are also described, because these theories are valid only when certain boundary or initial conditions are met.

2.1.1 Infinite Slope Theory

The infinite slope theory assumes that the failure plain is parallel to the surface of the sedimentary slope (Fig. 2.1) and that the edge effects (where this assumption is not valid) are insignificant. Further, the model assumes that full shearing resistance will be mobilized across the failure plain at a limit equilibrium (i.e., when the sediment mass is just at the point of failing). When using the limit equilibrium method (evaluation of forces of a system at the moment of failing) in a slope stability analysis, it must be recognised that there are a number of factors that cannot be accounted for, but may have a significant effect on the sediment mass stability, including discontinuities in the sediment, variation of the sediment properties with depth, variable pore pressures, stress deformation characteristics and the initial stress level. Sampling limitations (to estimate the relevant parameters) must also be considered when evaluating the results of a stability analysis (Busch & Keller, 1983). However, most workers studying seafloor slopes have used the infinite-slope simplification as the basis for stability analysis (e.g., Morgenstern, 1967; Almagor & Wiseman, 1978; Hampton *et al.*, 1978; Booth, 1979; Prior *et al.*, 1979; Busch & Keller, 1983; Booth *et al.*, 1985; Lee & Edwards, 1986; Nitzsche, 1989; Kostaschuk & McCann, 1989; Lee *et al.*, 1991; Merifield, 1992; Lee *et al.*, 1993; Roberts & Cramp, 1996). This is because the observed failures have generally so large areas that the infinite-slope assumption is nearly met, and the mechanical parameters are anyway not known well enough to justify greater sophistication (Lee & Edwards, 1986).

Mathematically, the model is an expression of balance between the maximum resisting force (the resisting force of the sediment just before the failure) and the shearing force. The ratio between the two forces is called the slope's factor of safety (FS), specifying the chances of a failure:

$$FS = \frac{F_{r,max}}{F_s} \qquad (2.1)$$

This is an estimate of the ratio of the maximum resisting forces to the shearing forces, where $FS > 1$ indicates stability, $FS < 1$ indicates instability, and $FS = 1$ indicates limit equilibrium.

The shearing force (F_s) is the downslope component of the normal force, or the effective weight of the sediment. Assuming a unit length of the potential sliding block of sediment, it is clear from Figure 2.1 that the effective shearing force can be expressed as:

$$F_s = \gamma_b z \sin \alpha \qquad (2.2)$$

where γ_b is the sediment's buoyant unit weight (i.e., the "submerged" sediment weight per volume, $\frac{(\rho_{sediment} - \rho_{water})Vg}{V}$, where ρ is density, V is volume, and g is the acceleration due to gravity, $\approx 9.81 \frac{m}{s^2}$), z is the thickness of the potential sliding block of sediment, and α is the slope angle.

The strength of a granular material against the stress that is acting to produce internal deformation, or shear, may derive from a number of factors (Carson & Kirkby, 1972). One is the plane friction produced when one grain attempts to slide past another. A second is the interlocking of the particles that constitute the material. These two factors are often considered jointly as the internal friction of the material, with the frictional resistance of the granular material usually expressed as the angle of internal friction, ϕ, (or the angle of shearing resistance (see section 2.1.2)). The tangent of this angle is termed the coefficient of internal friction. The total force developed by this frictional strength is the product of the internal friction coefficient of the material and the force that acts to push the particles together. This force is termed the effective normal force, or, as stress, the effective normal stress. In addition to the frictional strength, there is an electrostatic force that acts to bind the electrically-charged fine particles, and this second major source of shear strength is called cohesion. Its magnitude is neglible in sand and coarser silt, but highly significant in clay and clayey mud (W. Nemec, pers. commun., 1998).

17

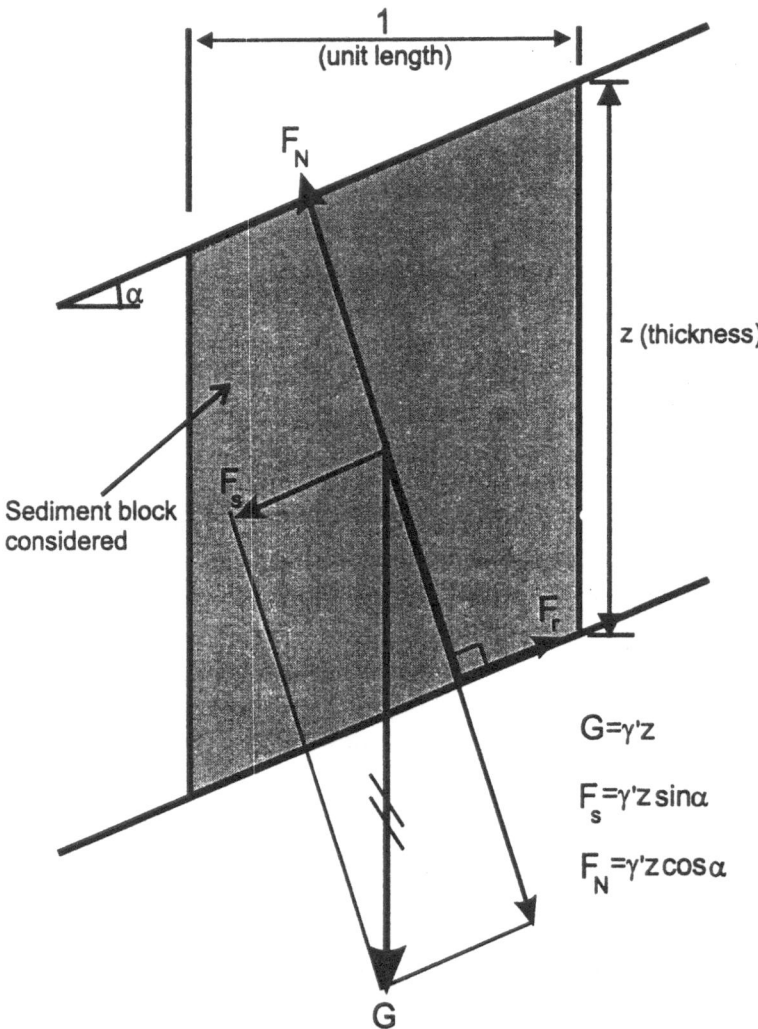

Figure 2.1: Infinite slope model for slope-stability analysis. Symbols: α = slope angle for the assumed failure plain (parallell to the surface of the sediment), z = thickness of the sediment, G = buoyant weight of the sediment, F_N = normal force, F_s = shearing force, and F_r = resisting force.

It is relatively easy to consider the shear strength in terms of the four elementary components (plane friction, interlocking, effective normal force, and cohesion), but rather difficult to picture the exact manner in which they combine. The most common and simplest interpretation of the shear strength (S) is thus the classic equation given by Coulomb (1776):

$$S = c + \sigma \tan \phi \qquad (2.3)$$

where c is the cohesive strength component, and the product of the normal stress, σ, and the angle of internal friction, ϕ, is the frictional strength component. In terms of the effective stress (see section 2.1.2), the equation becomes:

$$S = c\prime + (\sigma - u) \tan \phi\prime \qquad (2.4)$$

where $c\prime$ is cohesion with respect to effective stress, u is the pore pressure, $(\sigma - u) = \sigma\prime$ is effective normal stress, and $\phi\prime$ is the angle of internal friction with respect to effective stress (Terzaghi, 1962). The adjective 'effective' is applied to cohesion and angle of internal friction because natural sediment do not have a linear strength envelope as given by Equation 2.3, but have a concave-downward envelope. Inserting for the effective normal stress $(\sigma\prime)$ and cancelling out the pore-pressure term, the equation becomes:

$$S = c\prime + (\gamma_b z \cos^2 \alpha - \Delta u) \tan \phi\prime \qquad (2.5)$$

where Δu equals the pore-water pressure in excess of the hydrostatic pressure (see section 2.2). Assuming unit sediment block width, the basal area of the sediment block in Figure 2.1 is equal to $\frac{1}{\cos \alpha}$ (or $\sec \alpha$). Since stress is defined as force per area, the maximum resisting force can be expressed as:

$$F_{r,max} = [c\prime + (\gamma_b z \cos^2 \alpha - \Delta u) \tan \phi\prime] \frac{1}{\cos \alpha} \qquad (2.6)$$

Now, from equation 2.1, FS can be expressed as:

$$FS = \frac{[c\prime + (\gamma_b z \cos^2 \alpha - \Delta u) \tan \phi\prime] \frac{1}{\cos \alpha}}{\gamma_b z \sin \alpha} =$$

$$\frac{c\prime}{\gamma_b z \sin\alpha \cos\alpha} + \frac{(\gamma_b z \cos^2\alpha - \Delta u)\tan\phi\prime}{\gamma_b z \sin\alpha \cos\alpha} =$$

$$\frac{c\prime}{\gamma_b z \sin\alpha \cos\alpha} + \frac{\left(1 - \frac{\Delta u}{\gamma_b z \cos^2\alpha}\right)\tan\phi\prime}{\tan\alpha},$$

which gives:

$$FS = \frac{c\prime}{\gamma_b z \sin\alpha \cos\alpha} + \left(1 - \frac{\Delta u}{\gamma_b z \cos^2\alpha}\right)\frac{\tan\phi\prime}{\tan\alpha} \qquad (2.7)$$

Hampton *et al.* (1978), have attempted to account for ground accelerations due to earthquakes, by modifying the basic infinite-slope equation (2.7) through adding the terms for vertical and horizontal acceleration. The modified eqation (with $c\prime$ assumed to be zero) is:

$$FS = \left[\frac{1 - (\frac{\gamma}{\gamma_b})a_y - (\frac{\gamma}{\gamma_b})a_x \tan\alpha - \frac{\frac{\Delta u}{\gamma_b z}}{\cos^2\alpha}}{1 - (\frac{\gamma}{\gamma_b})a_y + (\frac{\gamma}{\gamma_b})\frac{a_x}{\tan\alpha}}\right]\frac{\tan\phi\prime}{\tan\alpha} \qquad (2.8)$$

where γ and γ_b are the total and buoyant unit weights of the sediment, and a_x and a_y are the coefficients of the horizontal and vertical acceleration (expressed in terms of the gravitational acceleration).

One of the effects of ocean waves are cyclic changes of the pore-water pressure in the seafloor sediment, and this loading pressure exerted on the sediment surface is thus of particular interest to the slope-stability problems. As a wave passes, an increase in the pressure at the sea bottom occurs below the wave's crest, whereas beneath the wave's trough there is a pressure decrease. The magnitude of these pressure changes, which are in phase with the wave, depends upon the wave length, the water depth and the wave height (Henkel, 1970). The differential loading of the seafloor will impose cyclic stress on the underlying sediment; and the resulting pore-pressure rises tend to accumulate in a low-permeability sediment. The stress reverses its direction with each half-cycle due to the periodic nature of waves and the underlying sediment is thus subject to cyclic stress reversals. If the stress exceeds the strength of the sediment, a mass failure may occur. This leads to a remoulding of the sediment, with a consequent reduction in the sediment strength and dissipation of the excess pore-water pressure (Henkel, 1970). For ocean-wave loading, Seed & Rahman (1978) gave an equation that specifies the induced normalized shear stress at the surface of a flat elastic seafloor. The equation, modified to include the influence of gravity on a seafloor slope, is:

$$\frac{\tau_w}{\gamma_b h} = \sin \alpha + \frac{\pi \gamma_W H}{\gamma_b L \cosh \left(\frac{2\pi W}{L} \right)} \tag{2.9}$$

in which τ_w = the wave-induced shear stress; γ_b = the buoyant unith weight of the sediment; h = the subbottom depth; α = the slope angle; γ_W = the unit weight of water; H = the wave height; L = the wave length; and W = the water depth. This shear stress comes in adittional to the shear stress produced by the gravity alone.

2.1.2   More About Shear Strength

As mentioned in the previous section, the strength of a sediment derives from four factors. These are reviewed here in more detail, with the discussion of the first three of them based on Carson & Kirkby (1972).

Plane Friction

Imagine a situation in which two dry slabs of sandstone, with smooth faces, rest upon each other, and the lower slab is progressively tilted until the upper slab just begins to move (Fig. 2.2). The forces acting on the upper slab are depicted in the latter diagram. The weight of the upper slab is divided into $G \sin \theta$, acting along the slope, and $G \cos \theta$ acting normal to the slope (where $G = mg$ is the weight of the upper slab and θ is the critical angle for the upper slab sliding). The latter force is countered by force F_N acting in the opposite direction and equal in magnitude. The shear force (F_s) is balanced by the frictional force (F_r) acting upslope at the contact of the two slabs. The shear strength, which in this case is the friction between the two slabs, depends upon the coefficient of plane friction (f) and the

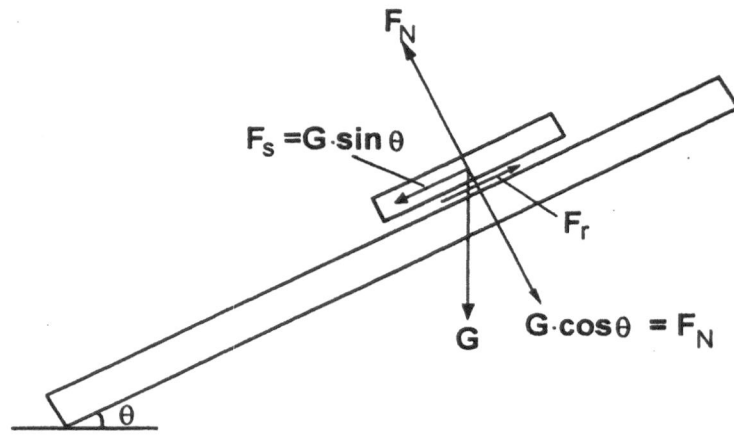

Figure 2.2: Demonstration of angle(s) of plane friction. θ = critical angle at which the upper slab begins to slide, G = weight of the upper slab, F_N = normal force, F_s = shearing force at the slab interface, F_r = resisting force.

weight force acting to press the two slabs together. In general terms:

$$F_r \leq F_N f \qquad (2.10)$$

where the two sides of the expression are equal at the time when the upper slab is just about to slide (limit equilibrium). At that moment, naturally, the shear force and the shear strength are also just balanced, which means:

$$F_s = F_r$$

$$G \sin \theta = G \cos \theta f$$

giving

$$\tan \theta = f \qquad (2.11)$$

The tilting of the slabs until the upper one begins to slide is thus one way to determine the coefficient of friction relative to a surface (or rather an interface). Strictly speaking, this coefficient refers to the starting or static friction only; once the sliding has begun, the friction decreases slightly to a new value, known as the coefficient of sliding (dynamic or kinetic) friction. This could be demonstrated in the above experiment (Fig. 2.2) by slowly decreasing the tilt of the slabs, during the sliding, beneath the initial angle θ. The upper slab would slide and accelerate, but at a decreased angle; the acceleration would cease and the slab would continue to slide at a constant velocity. At this time, assuming no air resistance, the shear force and the friction force are again just balanced, and application of the Newtonian Second Law of Motion yields, once again, equation 2.11. Since angle θ is now lower than at the start of the sliding, it follows that the angle of sliding friction (or the angle of rest; see section 5.3.1) is lower than the angle of static friction (or the critical angle of repose; see section 5.3.1). It should be emphasized that the actual friction coefficients for an interface depend considerably upon the contact characteristics (roughness, area, etc.) and many other factors. This aspect has been discussed in some detail by Horn & Deere (1962).

Interlocking

Tests similar to that just described (Fig. 2.2) for the determination of the friction coefficients of flat sedimentary intersurfaces can also be employed to measure the shear strength of a granular material (a directshear apparatus, such as that shown in Fig. 2.3, is widely used by civil engineers). In this situation, the resistance against shear derives from the interlocking of grains *in addition* to the plane friction along an interface. For the upper box and its sediment content to slide relative to the lower sediment-filled box (Fig. 2.3), it is necessary that some particles move upwards over the others. The amount of work necessary to perform this movement will increase with the magnitude of the load applied (Fig. 2.3). Both the interlocking strength, and the plane friction increase with the increasing effective normal stress. The test, which gives the value for the angle of internal friction (see Fig. 2.4), will first show the highest strength (ϕ_{max}) and then its decrease to an ultimate residual value ($\phi_{residual}$), which resembles the change from the high static friction to a lower sliding friction in the test shown in Fig. 2.2. This analogy, however, is only apparent (Carson & Kirkby, 1972); by the time the peak shear strength of the material has been attained, it is very

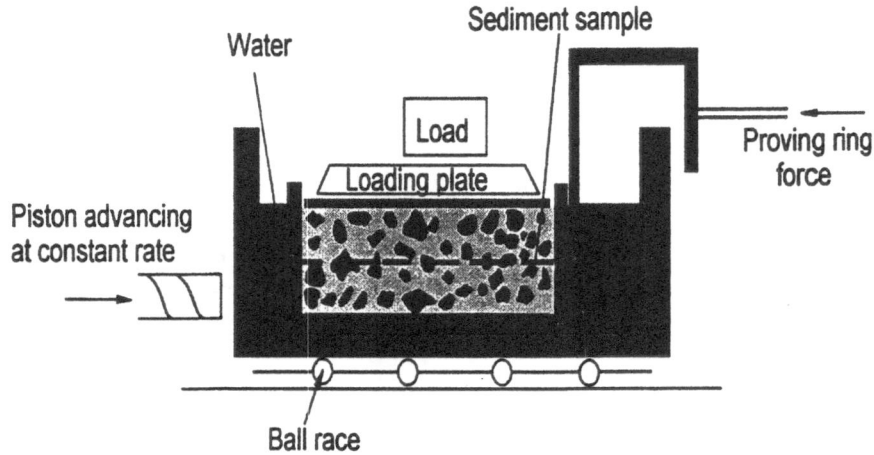

Figure 2.3: The direct-shear apparatus (modified from Carson & Kirkby, 1972).

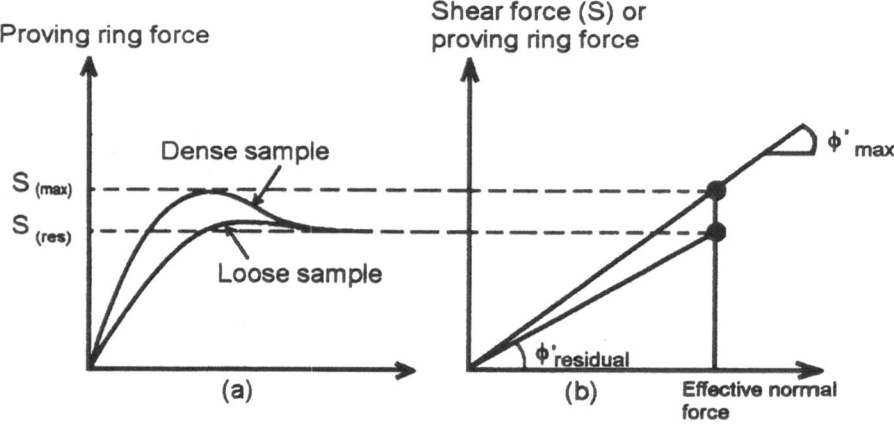

Figure 2.4: Results of a typical direct-shear test (modified from Carson & Kirkby, 1972): (a) change in strength of one sediment sample during a single test; (b) summary of many tests performed at different effective normal forces.

probable that true interparticle friction would have already dropped to the dynamic, sliding value. The relationships between the static and dynamic coefficients of plane friction and the peak and dynamic frictional strengths are only partly understood by engineers. In addition, the peak and residual strengths are seldom differentiated in the litterature. It is also worth noting (Fig. 2.4) that the deviation between the two angles is very small for loose sediment. Consequently, this issue is not further discussed here, apart from stating that the peak strength (or the critical-state friction angle; Schanz & Vermeer, 1996), if measured, is the correct parameter to use in prefailure stability analysis for it renders the assumption of full shearing resistance valid in the infinite slope theory. As point out by Schanz & Vermeer (1996), "the strength of sand is usually characterized by the ... critical state friction angle".

The angle of internal friction is dependent upon grain packing, mineral composition of the grains and the state of their surface chemistry, and the roundness, or angularity of the grains (Frossard, 1979; Murray, 1994). High values of the angle of internal friction have generally been attributed to the cementation and interlocking of carbonate particles at high strains (Nacci *et al.*, 1974), or to the formation of clay-organic aggregates in mud-rich sediments (Busch & Keller, 1983). Adams (1965) obtained $\phi\prime$ values as high as 48° from samples of amorphous peat. Varying clay mineralogy may also influence the variation of the friction angle. For example, Olson (1974) has demonstrated that $\phi\prime$ decreases in the sequence kaolinite $\rightarrow$ illite $\rightarrow$ smectite. Low values of $\phi\prime$ can be attributed to strong deformation. In a sediment with clay content greater than approximately 35 vol. %, the strain can induce preferred orientation of the clay particles in the shear zone and cause a reduction of $\phi\prime$ (Skempton, 1964). Angles of internal friction as low as 10° are not uncommon in clays that have been subject to large strains (Morgenstern, 1967).

Effective Normal Stress

The total normal stress is defined as the force acting normal to the shear surface per unit area of that surface (Carson & Kirkby, 1972). This stress is absorbed by the sediment in two different ways. Some of it is taken up by the grain contact along the shear surface, and some is taken up by the pore fluid. The total normal stress σ is thus given approximately by:

$$\sigma = \sigma\prime + u \qquad (2.12)$$

where $\sigma\prime$ is the effective normal stress (the inter-particle force per unit area of the shear surface) and u is the pore pressure. In the development of frictional resistance, it is clearly the effective normal stress, not the total stress, that is important. In situations where the effective normal stress is employed in equation 2.3, it is a convention to denote cohesion and internal friction as $c\prime$ and $\phi\prime$, respectively (cf. equation 2.4).

The reason why equation 2.12 is only approximately true has been summarized by Wu (1966), but the discrepancy in reality is so slight that it can be ignored here. As indicated in equation 2.12, the pore pressure, due to its control on the effective normal stress, is thus extremely important in determining the amount of internal friction involved.

Cohesion

In general terms, a sediment is considered to be cohesive if the particles adhere to one another after wetting and subsequent drying and if significant force is then required to crumble the sediment (this does not include sediments whose particles when wet, adhere due to surface tension) (Craig, 1987). If the sediment grains are easily moved as individuals, the sediment is said to be non-cohesive. Cementation, electrostatic and electromagnetic attractions, and primary valence bonding and adhesion have been considered as the sources of cohesion. Capillary stresses and micro-frictional forces may be a source of the so-called apparent cohesion, which does not involve inter-particle bonding (Mitchell, 1993).

Cementation is a chemical bonding between particles, which can develop as a result of the inter-particle precipitation of carbonate, silica, alumina, iron oxide and/or organic compounds. The cementing mineral may be derived from the sediment itself, as a result of solution-precipitation process, or may be brought in solution from the exterior. Cohesive strength as high as several hundreds kPa may result from cementation.

Electrostatic and electromagnetic forces include the attraction between clay particle edges with opposite signs (charges) and the van der Waals attractions between closely spaced clay particles. The electrostatic attraction becomes significant for separation distances < 2.5 nm, whereas the electromagnetic or van der Waals forces are a source of tensile strength only between closely spaced particles of very small size ($< 1\,\mu$m).

Primary valence bonding and adhesion may play some role, too. When a normally consolidated clay is unloaded, thus becoming overconsolidated, the strength does not decrease in proportion to the effective stress reduction; instead, part of it is retained. Wheter or not the higher strength in the overconsolidated clay is due to the lower void ratio or the formation of interparticle bonds is not known. However, a "cold welding", or adhesion, may be responsibe for some of it. This could result from the formation of primary valence bonds at interparticle contacts.

Capillary stresses result from the combination of water attraction to sediment particle surfaces and the water property of surface tension, which cause an apparent attraction between particles in a partly saturated sediment. This is not a true cohesion, but rather a frictional strength generated by the positive effective stress due to the negative pore-water pressure.

Apparent mechanical forces, or micro-friction, are du to the particle shape and packing, which may cause an apparent cohesion in a sediment with no physical or chemical attractions between particles.

The cohesion term in equations 2.3 and 2.4 involves only true cohesion; apparent cohesion is accounted for by the frictional parameter in the same equation.

It is well-known that even a small cohesion is sufficient to support a thick sedimentary column (Almagor & Wiseman, 1978), and the importance of cohesion in sedimentary slopes was emphasized by Gray & Leiser (1982), who noted that "not much" cohesion was required to maintain a stable slope.

2.1.3 Consolidation

Equation 2.7 contains a term called "excess pore pressure" (discussed further in section 2.2). This pressure can be calculated using Gibson's consolidation theory (discussed in section 2.1.4). In order to follow this theory, some aspects of the engineering meaning of the term consolidation need to be considered.

Consolidation is a time-dependent mechanical reduction of the sediment volume, usually by the loss of pore volume due to loading (Maltman, 1994). Consolidation is time-dependent simply because it takes time to dissipate the excess pore fluid that is generated as a consequence of the reduction of pore volume. The time it takes is determined by the permeability and pore tortuosity of the sediment and the viscosity of the pore fluid (which again is dependent upon chemistry and temperature). The total stress acts on both the grain framework (effective stress) and the pore fluid (pore-fluid pressure), but the instantaneous response of a saturated sediment to an increase in the total stress is an increase in the pore fluid pressure. This increased pressure makes the fluid flow from the sediment and thus causes a progressive decrease in the pore volume, whilst the effective stress increases until an equilibrium state of consolidation is re-established (Jones, 1994). Hence, the process of consolidation continues until the excess pore-water pressure set up by the increase in total stress has completely dissipated (see Craig, 1987, Chapt. 5). In sand and other permeable sediments, the time required for a full consolidation is very short and the burial is not accompanied by the development of an excess pore-fluid pressure, unless the sand becomes sealed by the surrounding sediments of low permeability (Jones, 1994).

Consolidation can be divided in two stages. Primary consolidation, as discussed above, is dominated by the expulsion of the pore fluid, whereas secondary consolidation (also called creep) is dominated by fine-scale pore reduction involving minute adjustments in the grain framework. The possible causes of secondary consolidation include grain surface diffusion, time-dependent generation of cracks (micro-fractures) associated with the redistribution of sorted strain energy, and chemical diffusion in microfractures, with stress corrosion weakening the fracture tips (Jones, 1994).

Consolidation should not be confused with compaction. Compaction is permanent reduction in the bulk volume of a sediment, including adjustments of *both* the grains and the pore space between them. In other words, consolidation is an aspect of compaction.

Basically, three states of consolidation can be distinguished. Underconsolidation is a state in which the sedimentation occurs so rapidly that the pore-water pressure set up by the loading does not fully dissipate. Rather, the pore pressure gradually increases and progressively supports more of the weight of the overlying sediment particles (Schwab *et al.*, 1987). The sediment thus becomes overpressured (i.e., develops an excess pore pressure) and has a more open framework, with a higher water content, than it would have under normal consolidation (Collins & McGown, 1974). Consequently, the sediment has a lower shear strength. If the effective stress had at some time been greater and the sediment were at that time normally consolidated, it would be overconsolidated (Craig, 1987). This might be a result of the erosion of overburden (e.g., melting of an ice-sheet after glaciation). An overconsolidated sediment has a higher shear strength than a normally consolidated one. If the current effective stress is the maximal to which the sediment has ever been subjected and the sediment has no excess pore pressure, it is said to be normally consolidated.

The coefficient of consolidation (ς) comes from one-dimensional consolidation theory and relates the decrease in sediment area perpendicular to the stress with the pressure and time. It is defined by the equation:

$$\varsigma = \frac{\kappa}{\rho_W M} \qquad (2.13)$$

where κ = permeability, ρ_W = density of pore water, and $M = \frac{e_1 - e_2}{P_2 - P_1(1 + e_1)}$, where e is the void ratio (volume of voids relative to volume of grains) at pressure P. The coefficient is taken as a constant in the fundamental differential equation for one-dimensional consolidation (although it can change with the confining stress), and is thus limited by the basic assumptions of the theory. In addition to the difficulties that arise from the simplifying assumptions, the mechanical details of a consolidation test (see Been & Sills, 1981, for tests to determine the coefficient of consolidation) may radically affect the value of the coefficient (Leonards & Ramiah, 1960). This means that the coefficient in practice is difficult to determine with sufficient accuracy. In fact, tests to determine the value of this coefficient are seldom done, although some general considerations have been presented. Skempton & Bishop (1954) stated that the coefficient depends primarily on the size and nature of the sediment particles and is not influenced to any degree by the moisture content or the sediments mechanical state. Furthermore, Buchan *et al.* (1967) have noted that the coefficient shows a positive correlation with the sediment's sand fraction (%).

2.1.4   Gibson's Consolidation Theory

Gibson's (1958) consolidation theory is a mathematical theory describing the one-dimensional consolidation of a clay layer that increases in thickness with time. When the deposition is fairly uniform over an area whose dimensions are large compared with the thickness of the deposited layer, the consolidation will be approximately one-dimensional (Gibson, 1958). This theory can be used to estimate pore pressures in a sediment. The theory requires constant-rate deposition on an impermeable substratum base and further assumes constant bulk density, permeability, sediment compressibility and pore-water incompressibility. These are rather drastic assumptions, which may be acceptable for the modelling of the relatively thin, nearsurficial clay layers commonly encountered in civil engineering problems, but are hardly valid for thick and deeply burrowed clay layers (Audet & Fowler, 1992). However, the assumptions are not unreasonable for the basinal situations considered in stratigraphic modelling, and the theory is attractively simple to be used for the estimation of excess pore pressure in the slope stability analysis. Accepting the boundary conditions above, the excess pore pressure in a sediment is given as:

$$\Delta u = \gamma_b q t - \gamma_b (\pi \varsigma t)^{-\frac{1}{2}} e^{\frac{-h^2}{4\varsigma t}} \int_0^\infty \xi \tanh\left(\frac{q\xi}{2\varsigma}\right) \cosh\left(\frac{h\xi}{2\varsigma t}\right) e^{-\frac{\xi^2}{4\varsigma t}} d\xi \quad (2.14)$$

where γ_b is the buoyant unit weight, q is sedimentation rate, t is time, ς is coefficient of consolidation, h is height (where Δu is equated) above the impermeable substratum (basin floor), $Z(t)$ is the total sediment thickness at time t, and $\xi = \frac{h}{Z(t)}$.

It is to be emphasized that equation 2.14 applies to the excess pore pressure that may arise as a direct consequence of deposition. Excess pore pressures due either to the shear stress set-up within the slope or to a general dynamic loading are not considered.

Examples of studies that have used both the infinite slope theory and the Gibson consolidation theory to analyze the stability in subaqueous sedimentary slopes include Morgenstern (1967), Almagor & Wiseman (1978), Hampton *et al.* (1978), Busch & Keller (1983), Booth *et al.* (1985) and Nitzsche (1989). However, these authors have used the consolidation theory only in a 'graphical' way, by employing a diagram given in Gibson's paper to estimate indirectly the value of excess pore pressure; no numerically solution of the equation has been given in any of the papers.

2.2 Slope Failure

Slope failure refers to the process that causes sliding (i.e., a slide is a direct result of slope failure). Unstable sediment on a subaqueous slope can fail and further evolve into sediment gravity flows, thus being subject to resedimentation. With the theoretical background given in section 2.1, it is now possible to consider the factors and processes leading to submarine slope instability (as summarized in Fig. 2.5).

As already mentioned (Chapter 2), the failure of a sedimentary slope occurs when the gravitational shear stress exceeds the sediment's shear strength. In other words, an increase in the shear stress or a decrease in the shear strength can lead to a slope failure. On the basis on equations 2.2 and 2.4, a discussion of the factors that can lower the strength or increase the stress, is given below. Further, the geological processes that may lead to similar changes are discussed.

As is clear from equations 2.2 and 2.4, a change in one or more of the following parameters will affect the shear stress and/or the shear strength:

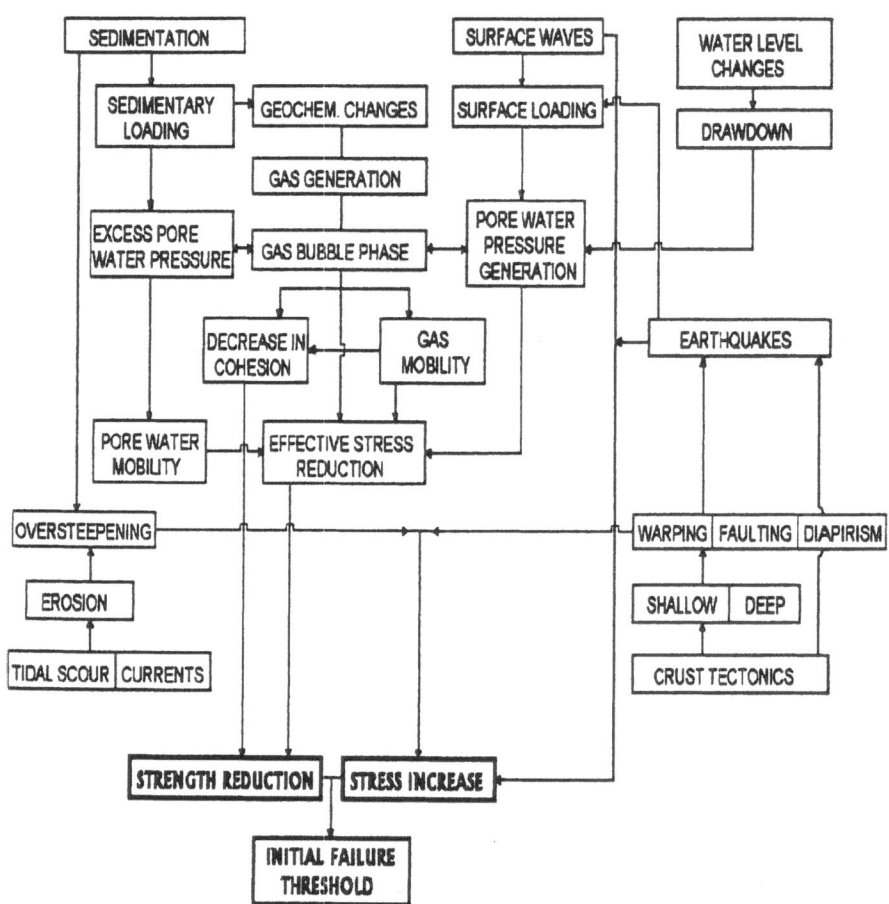

Figure 2.5: Relationships between the mechanical factors and processes responsible for submarine slope instability (modified from Prior & Coleman, 1984).

Depositional Angle. – The greater the depositional angle of the sediment the greater the gravitational shear stress (see equation 2.2, Fig. 2.1).

Sediment Mass. – The greater the sediment mass, the greater the shear stress (Fig. 2.1), although a increase in the mass will also affect the resisting force (see equation 2.6). For slopes of low and moderate inclination, the shear resistance (which is proportional to the normal force, F_N, and increases with the cosine of the slope angle) will grow more rapidly with the increasing burial thickness than the shear stress will (which increases with the sine of the slope angle), although with steeper slopes (for which the sine value increases more rapidly than the cosine) the opposite will be true. Hence, the sediment on a slope above a certain inclination will become increasingly unstable with the progressive burial. In addition, when cohesion (section 2.1.2) is involved, the increase in the strength is much lower than the increase in stress, simply because only the frictional part ($\phi\prime$) is related to the effective normal stress, whereas the cohesion element ($c\prime$) is constant with depth.

Pore Pressure. – If, after a given increment of loading, the sediment lacks sufficient permeability for the pore fluid to escape, the latter, by being incompressible, has to sustain a disproportionate part of the load (Gretener, 1981). This pressure in excess of the normal fluid pressure is termed the excess pore-fluid pressure, and is associated with unconsolidated or undrained sediments. Normal fluid pressure derives from that portion of the burial load arising from the overlying column of pore fluid, plus the overlying sea-water column, if present. The effect of the excess pore pressure in a sediment is the reduction in the effective stress, and hence of the sediment strength (see equation 2.4). It is likely, in fact, that most failures of sedimentary slopes occur due to overpressuring (Maltman, 1994).

The most important geological processes that may lead to a change in the above-mentioned parameters are:

Sedimentation Rate. – The higher the sedimentation rate, the greater is the mass of the sediment deposited, and thus the greater the resulting gravitational shear stress (although this latter also leads to a higher sediment strength, as discussed earlier). High sedimentation rate can also lead to the buildup of excess pore pressure in the deposit, especially in sediment with a low permeability.

Changes in Relative Sea-Level. – A relative sea-level fall may activate the excess pore pressure in the sediment, by reducing the hydrostatic confining pressure, and reduce the sediment's strength (e.g. Lee *et al.*, 1996). Both the sedimentation rate and tectonics together with eustasy and compaction, are the main controls on the depositional angle in sediment, thus determining as to whether an oversteepening of the slope will occur.

Cyclic Loading. – This factor increases the pore pressure in the sediment and results from the transmission of oscillatory seismic waves, and in subaqueous settings no deeper than about 200 m, from the passage of surficial sea waves (Seed & Rahman, 1978).

Gas Generation. – The generation of interstitial gas leads to an increased pore pressure in the sediment, and thus reduces the shear strength of the latter.

Other geological processes that are thought to be of minor importance to the subaqueous sediment stability are creep deformation and the breakage of interparticle bonds, as well as several processes known to produce excess pore pressure in sediment, such as mineral dehydration (Colten-Bradley, 1987), biopressuring resulting from organic decay (Nelson & Lindsley-Griffin, 1987), geothermal heating (Barker, 1972; Magara, 1975; Barker & Horsfield, 1982; Daines, 1982), migration and gravitational segregation of multiphase fluids in gas-oil-brine systems (Fertl, 1973; Hedberg, 1974; Archer & Wall, 1986; Young & Low, 1965) and osmosis (Hanshaw & Zen, 1965; Young & Low, 1965).

Chapter 3

Postfailure Evolution of Sediment Mass Movement

Not every slope failure necessarily leads to a massflow (i.e., every slope failure leads to a slide, but this will not necessarily evolve into a sediment gravity flow). Whitman (1985) has distinguished disintegrative and nondisintegrative failures. In a nondisintegtative failure, the deformation occurs during a transient loading event, but little strain continues after the event. In a disintegrative failure, in contrast, the transient loading event produces sufficient loss of shear strength that a sediment gravity flow is generated. Poulos *et al.* (1985) developed the notion of a steady state of deformation as a way of quantifying the development of a disintegrative failure. The steady-state approach, which is based on extensive laboratory studies and follows the work of Casagrande (1936), is consistent with the concepts of critical-state soil mechanics (Schofield & Wroth, 1968). The initial (prefailure) void ratio and effective-stress conditions can be plotted on a steady-state diagram to assess whether a disintegrative failure of the sediment is possible (Fig. 3.1). If the projection points of the initial-state data fall *above* the steady-state line and the loading is undrained, the pore pressure will increase and the effective stress will decrease in the sediment during failure. The steady-state shear strength under the movement will be lower than the initial-state shear strength, and the first criterion for mass flow is thus satisfied. Beacause the sediment grain fabric tends to collapse due to the shear, the sediment behaviour is called "contractive". Accord-

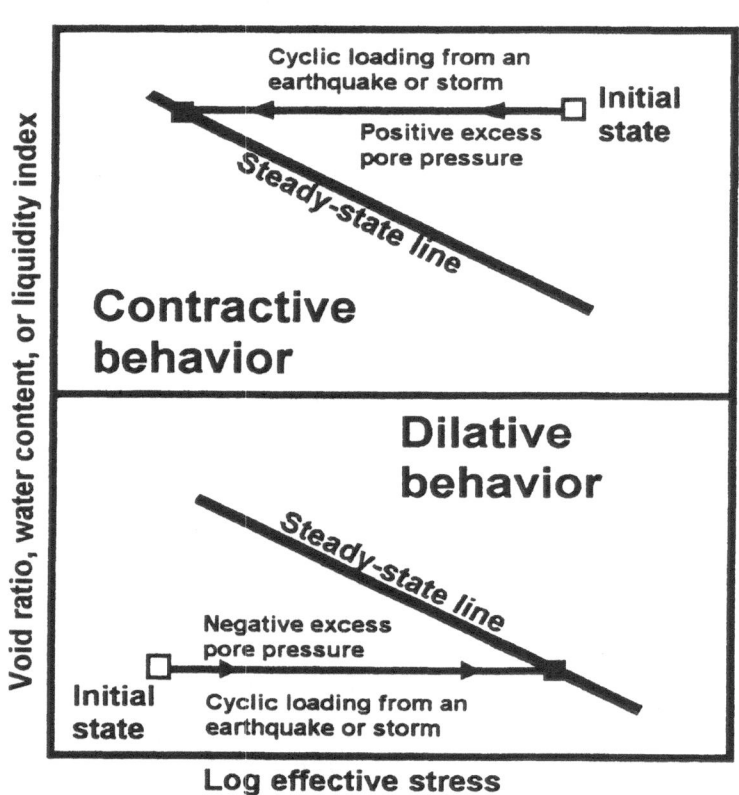

Figure 3.1: Two ways in which the state of a sediment can change during an undrained failure caused by a transient loading event such as an earthquake or a passage of storm waves (modified from Hampton *et al.*, 1996).

ing to Whitman (1985) and Poulos *et al.* (1985), only a contractive sediment can result in a disintegrative failure. But, if the projection points of the initial-state data plot *below* the steady-state line, the sediment will tend to dilate and decrease its pore pressure. The "dilative" sediment behaviour cannot produce a disintegrative failure, unless the grain fabric somehow expands and is converted to a contractive state. Sands and normally consolidated clays often show a contractive behaviour, whereas dense sands and overconsolidated clays more often show a dilative behaviour (Hardin, 1989; Shimizu, 1982). In summary, if a loss of shear strength occurs in the sediment mass during its failure, such that the resulting strength is smaller than the downslope gravitational stress, the failing sediment mass will tend to accelerate and turn into a massflow.

3.1 Sediment Gravity Flows

A sediment gravity flow, also referred to as massflow, is defined as a sediment mass movement in which the sediment mass has been pervasively sheared (remoulded) and its motion parallel to the substratum is directly due to the pull of gravity. Five types of massflows can be distinguished on the basis of how the grains are supported above the floor (Middleton & Hampton, 1976; Lowe, 1982; Nemec & Steel, 1984), namely: turbidity currents, fluidized sediment flows, liquified sediment flows, cohesionless debrisflows, and cohesive debrisflows (Fig 3.2).

Sediment gravity flows are distinguished from sediment slides or slumps on the basis of the degree of internal deformation, which is pervasive in flows, internal but discrete in slumps, and limited to the basal surface in slides (Middleton & Hampton, 1976; Nemec, 1990).

Liqified and fluidized sediment flows have the sediment supported by the escaping fluid – partially in the former case (where the fluid escapes only because the grains settle, displacing it) and fully in the latter case (where the pores are supplied with excessive fluid volumes). These flows are thougth to be of little importance in the transport of sediment over long distances on the basin floor, compared to turbidity currents and debris flows. Fluidized flows are chiefly pyroclastic, or require some very special flow conditions,

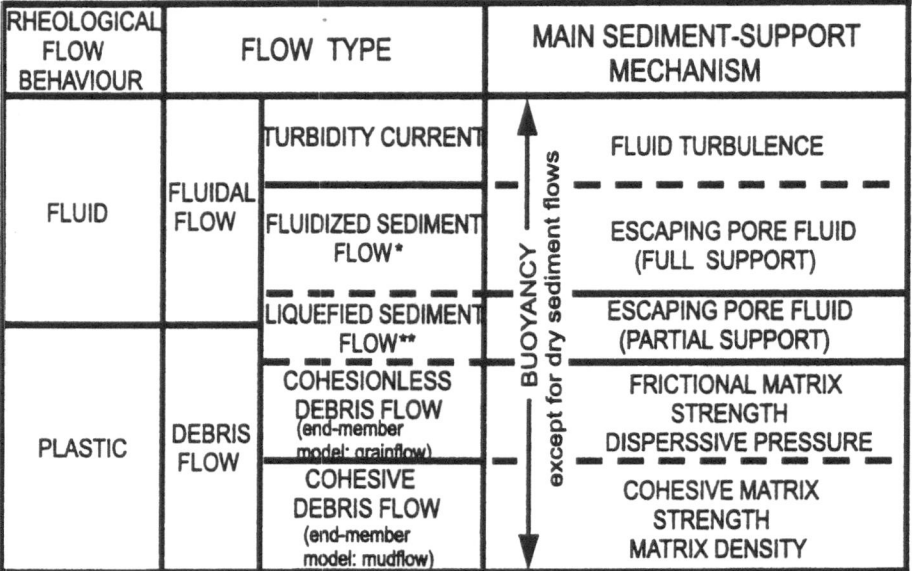

RHEOLOGICAL FLOW BEHAVIOUR	FLOW TYPE		MAIN SEDIMENT-SUPPORT MECHANISM
FLUID	FLUIDAL FLOW	TURBIDITY CURRENT	FLUID TURBULENCE
		FLUIDIZED SEDIMENT FLOW*	ESCAPING PORE FLUID (FULL SUPPORT)
		LIQUEFIED SEDIMENT FLOW**	ESCAPING PORE FLUID (PARTIAL SUPPORT)
PLASTIC	DEBRIS FLOW	COHESIONLESS DEBRIS FLOW (end-member model: grainflow)	FRICTIONAL MATRIX STRENGTH DISPERSSIVE PRESSURE
		COHESIVE DEBRIS FLOW (end-member model: mudflow)	COHESIVE MATRIX STRENGTH MATRIX DENSITY

BUOYANCY except for dry sediment flows

Figure 3.2: Classification of sediment gravity flows (from W. Nemec, pers. comm., 1996; based on Middleton & Hampton (1976); Lowe (1982); Nemec & Steel (1984).

*Chiefly pyroclastic flows, with gas generated from within or injected along the base.

**Transient state; stops in minutes or seconds, or accelerates and turns into either a debrisflow or a turbidity current.

where abundant fluid is continually streaming through the moving sediment mass – from its base upwards. A liquefaction often occurs at the starting phase of a massflow, or during the early stages of deposition from a high-density turbidity current. Van der Knaap & Eijpe (1968) and Middleton (1969) have calculated that a 10-m thick layer of fine sand can be liquefied for only a few hours, and the time reduces to minutes and seconds for coarser or smaller thicknesses. This means that pure liquefied flows can move over relatively short distances only; or they either turn into debrisflows or accelerate and become turbidity currents. Grainflow is a type of cohesionless debris flow, where the sediment is supported by grain-to-grain collisions and the resulting dispersive pressure (Lowe, 1976). True grainflows composed of sand cannot exceed 5-10 cm in thickness, because the dispersive pressure generated by sand grains is insufficient to maintain the movement and support a thicker grain dispersion (Lowe, 1976). This can readily be seen, for example, on the slopes of aeolian dunes – where thickest grainflows are no more than a few centimetres thick, despite the very steep slopes. Thicker flows must have one or several additional grain-support mechanisms. Gravelly grainflows may be thicker, but require very steep slopes, in excess of 35° (Middleton & Southard, 1984).

Turbidity currents and debrisflows are common and relatively mobile, capable of spreading over large distances on a relatively gentle seafloor, and are generally considered to be of primary importance in submarine sedimentation (Stow, 1994). Turbidity currents are particularly important. For example, Boggs (1987) has emphasized that large, high-velocity turbidity currents generated on the shelf or upper slope are probably the single most important mechanism for the transport of sand and gravels into the deep water. Likewise, Walker (1980) has pointed out that turbidites are volumetrically the dominant slope and basin deep-water facies, which may constitute as much as 45 % of the total volume of all sedimentary rocks.

Therefore, the mechanical properties of turbidity currents and debrisflows are discussed here in some detail, although no effort is made to describe the internal characteristics of their deposits (such as the types of clast-size grading in debrisflow beds, or the Bouma divisions and other features of turbidites), simply because it is impossible to model such details in a dynamic slope model, whose resolution is different (see earlier section 1.2). For facies models and typical internal characteristics of sediment gravity-flow deposits the reader is referred to Allen (1984), Walker & James (1992) and Stow *et al.* (1996).

3.1.1 Turbidity Currents

A turbulent flow, as opposed to a laminar one, with roughly parallel stream-lines, is a flow with random motions superimposed upon unidirectional flow (Collinson & Thompson, 1989). The development of eddies and their breaking down into smaller vortices in a turbulent flow absorbs energy, whereby a greater shear stress is required to maintain a particular velocity gradient in a turbulent flow than in a laminar flow. While the shear stress (τ) in a laminar flow is expressed by $\tau = \mu(\frac{dU}{dW})$, the shear stress in a turbulent flow is expressed by $\tau = (\mu + \eta)\frac{d\overline{U}}{dW}$, where μ is the viscosity of the fluid, η is the so-called eddy viscosity (which is not a constant, but a variable dependent upon the scale and intensity of turbulence), U is the flow speed, $\overline{U}$ is its mean speed, and W is water depth (Collinson & Thompson, 1989). The turbulence developes in a flowing fluid when the flow's Reynolds number (Re) exceeds some critical value. The Reynolds number is a dimensionless parameter that expresses the ratio between the inertial forces (which are related to the scale and velocity of the flow and thus promote turbulence) and the viscous forces (which are related to the fluid's molecular friction and tend to surpress turbulence) (Collinson & Thompson, 1989).

A turbidity current is a type of turbulent density current, whose higher density, relative to that of the ambient water, is due to sediment suspension (Stow, 1994). Turbidity currents are often said to be driven by a mechanism called "autosuspension": the sediment load suspended by the current provides the energy (density), and the latter in turn allows the current to maintain or even increase the load. The sediment particles are lifted up by turbulence and the finer fraction is held in suspension due to the upward component of the turbulent eddies. The flow thus has a net upward flow flux, counteracted by the tendency of the sediment particles to settle under their own weight. In other words, the sediment and the flow are intimately coupled, because the particles are suspended within the current by the turbulence generated by the flow, and the flow as such is maintained by the negative buoyancy force set up by the particle suspension (Dade *et al.*, 1994). Since the current consists of a fluid and has a greater density than the ambient water, it is influenced of a greater gravity force than the latter and can flow on slopes of even less than 1°. The velocity of the current can be quite high, with the reported speeds of currents several hundred metres thick of the order of 20 m/s (Piper *et al.*, 1988), although the speed depends strongly on the current's density (see equations below).

Quantitative Aspects

A turbidity current, in reality, is a non-uniform, unsteady, non-linear and mainly free-boundary flow (Stow, 1994). With these characteristics it is not yet possible to construct a comprehensive theoretical model (Allen, 1985). Suffice it to say that turbulence alone, as a chaotic phenomenon, has this far defied numerical modelling. However, several important attempts have been made to explore certain quantitative aspects of turbidity currents. Geometrically, a turbidity current can be divided in the head, neck, body and tail segments (Middleton & Hampton, 1976). The coarsest material is carried in the head and the finest in the tail. Erosion from the head and deposition from the body and tail can occur at the same time (Stow, 1986). Middleton (1966a) has suggested a Chezy-type equation, in which the velocity of the current's head, U_H, is given by:

$$U_H^2 = 0.56 g d_H \left[\frac{\Delta \rho}{\rho} \right] \qquad (3.1)$$

where $\Delta \rho$ is the density difference between the current and the ambient water, ρ is the density of the latter, g is the acceleration due to gravity, and d_H is the thickness of the head. The average velocity of the current (or the speed of the current's body), $\overline{U}$, has been calculated by Middleton (1966b) as:

$$\overline{U} = \sqrt{\frac{8g}{f_0 + fi} \left(\frac{\Delta \rho}{\rho} \right) d_B \tan \alpha} \qquad (3.2)$$

where d_B is the thickness of the body, α is the substratum slope angle, f_0 is the dimensionless friction coefficient at the base of the flow, and f_n is the dimensionless friction coefficient at the top of the flow. One practical problem with using this equation is the lack of empirical knowledge of the friction coefficients.

The ratio of the head velocity to the body velocity is approximately 1.0 on gentle slopes, but less than 1.0 on steeper slopes.

The results above are based on laboratory experiments. Natural turbidity currents in deep-water basins are several orders of magnitude larger than those produced in laboratory flumes, so that the extent to which experimental results can apply to natural turbidity currents is somewhat disputable (Stow, 1986).

Flow Initiation

Turbidity currents are considered to be initiated in one of the following ways (Stow, 1986): (1) from the transformation of slides, evolving into slumps and increasingly turbulent debrisflows as they incorporate more water ; (2) from the sand masses supplied as suspension spillovers, grainflows or rip-current splays fed across the shelf into the heads of submarine canyons; (3) by storm stirring of unconsolidated bottom sediments and the build-up of a concentrated shelfal nepheloid bottom layer; (4) directly from the sediment suspension delivered to the basin by river floods or glacial meltwater outflow.

Incidence

The time frequency with which turbidity currents are generated and turbidites are emplaced in any particular part in the deep-water basin depends on such factors as the nature of the source area, the distance of the depositional area to the source, the seismicity of the source area and the relative sea-level stand (Stow, 1986). Delta-generated turbidity currents, produced during periods of high river discharges, may occur as often as once every two years (Heezen & Hollister, 1971). These turbidity currents are mainly low-density flows. In the proximal parts of active deep-sea fans, turbidites may be emplaced once every 10 years (Gorsline & Emery, 1959). However, the more distal fan slopes or remote basinplain environments

42

are draped with a turbidite every 1000-3000 years, although this frequency may vary a great deal (Rupke & Stanley, 1974; Kelts & Arthur, 1981; Stow, 1984).

Types of Turbidity Currents

Based on the initiation mechanism, turbidity currents can be divided into sustained and surge-type currents (Nemec, 1990).

Sustained turbidity currents develop when a turbulent, sediment-laden flow enters the basin, plunges on its marginal slope and continues to spread in the basin so long as the feeding occurs; when the turbulence is formed as a result of the density difference between the bottom water and the overlying water caused by the continual supply and settling of sediment suspension; or when a delta slope is subject to a long-duration retrogressive slumping due to the continual growth of river-mouth bars (Nemec, 1990; Lønne, 1995, 1997). This type of turbidity current is thought to be often related to a hyperpycnal or hypopycnal river effulent (Nemec, 1990). These currents can be channelized or not, and may penetrate far into the basin, depositing little or no sediment on the basin-margin slope, unless the latter is relatively wide and gentle. Sustained turbidity currents may have a duration from hours to days and can deposit sediment over vast basinal areas, although they mostly carry mud and silt, rather than sand (e.g., Chikita, 1990).

Surge-type turbidity currents are initiated by disintegrative slope failures (see the first part of this Chapter), and thus involve an early stage of slumping and debrisflow, during which the sediment mass incorporates water, shears and becomes turbulent. This triggering phase is relatively brief, and since the feeding involves a finite portion of sediment, the current peaks and wanes (Luthi, 1980). Processes leading to the slope failure and slumping have been discussed in section 2.2. According to Middleton & Hampton (1976), the turbidity currents in oceans are usually catastrophic surges, and these currents are by far the most important agent delivering coarse siliciclastic sediment to the deep-water basins.

Turbidity currents are able to move on very low-gradient slopes and may transport sediment of a wide range of grain sizes, from gravel to the finest clays, with a wide range of sediment concen-

43

trations. This latter aspect has an important implication for the mode of sediment deposition, and has led to the recognition of high- and low-density turbidity currents (Lowe, 1982). The point of this distinction is not the absolute density, which may vary from flow to flow depending upon the flow power (velocity and bottom shear stress); it is rather the mode of the deposition, particularly at the initial stage. Lowe (1982) has classified all turbidity currents that deposit the first or main part of sediment in a non-tractional fashion (i.e., by rapid dumping from turbulent suspension or as traction carpet) as high-density currents, and all turbidity currents that deposit sediment in a fully tractional fashion as low-density currents. Shanmugam & Moiola (1997), Shanmugam (1996) and Shanmugam *et al.* (1995) have recently suggested that the concept high-density turbidity currents should be discarded, because non-tractional deposition may mean a sandy debrisflow, but this view bears some misunderstanding and has been met with much critique (e.g., Slatt *et al.*, 1997; Lowe, 1997; Coleman Jr., 1997; Bouma *et al.*, 1997; D'Agostino & Jordan, 1997; Hiscott *et al.*, 1997). Luthi (1981) has shown that the density of a turbidity current is related to the flow velocity and can change within short distances, also due to the entrainment of the ambient water. Surge-type turbitity currents can have both high and low densities, whereas the sustained turbidity currents are more often of low density.

Autosuspension

It has been argued that a turbidity flow is self-maintained due to the mechanism called "autosuspension" (Bagnold, 1962; Pantin, 1979; Parker, 1982; Stacey, 1982; Stacey & Bowen, 1988). The concept of autosuspension
(Southard & Mackintosh, 1981) invokes a state of dynamic equilibrium in which: (1) the excess density of the suspended sediment propels the flow; (2) the flow is viscous and develops turbulence; (3) the turbulence keeps the sediment particles in suspension and possibly entrains more sediment; and so on. All that is needed to keep the loop intact is that the loss of energy to the viscosity (or the fluid's internal friction) must be compensated for by the gain in gravitational energy as the flow travels downslope and accelerates

(Stow, 1994). If the concept of autosuspension is valid, it would imply that a turbidity current can travel over an "infinite" distance without erosion and deposition, so long as the slope remains constant and the flow is inviscid. (An "inviscid" flow means one in which the frictional loss of energy to the bottom and the surrounding water, at the flow boundaries, is neglible). It has been argued (Southard & Mackintosh, 1981; Middleton & Southard, 1984), however, that true autosuspension is unlikely to occur, simply because the flow is not inviscid, looses energy, and some of the suspended sediment is inevitably deposited. Nevertheless, autosuspension can occur on shorter-term scale, and it is the steadiness of the flow density that is crucial to this "self-drive" phenomenon (Pantin, 1979).

Flow Dissipation and Deposition

The energy of a turbidity current is dissipated by friction at the boundaries and work against the fluid's viscosity, but may also be dissipated by channel-wall friction and the loss of mass by overbank flow in a submarine channel; furthermore, the spreading of the current, as it moves from the channel onto the basinal plain, rapidly reduces the flow thickness and the driving force of buoyancy and inertia (Middleton & Hampton, 1976; Normark, 1989). Even more importantly, the reduction of the slope gradient in basinward direction causes deceleration, which in turn reduces turbulence. All these factors cause reduction in flow momentum and deposition of the transported sediment.

The deposit of a turbidity current is called a turbidite (for detailed internal characteristics of turbidites, see Allen, 1984; Walker & James, 1992). The typical thicknesses of turbidites range from 0.1 to 10 m (Rothman et al., 1994), and the grain size ranges from mud to sand to gravel. Gosh et al. (1986) found, from the flume experiments with heterogeneous sediment mixtures, that all the grain sizes present in suspension were involved in the process of turbidite formation when the flow velocity decreased. In large natural currents, the coarsest material tends to fall out from the suspension first, giving rise to a normal grading (upward fining). The deposit, in a 2-D longitudinal cross-section, has a lens-shape geometry (lobate in 3-D terms). Luthi (1981) has found, in a large-scale laboratory experiment, that a rapid diluton of non-channelized turbidity current occurs, and that the thickness of the deposit and the mean grain size decrease radially away from the source. Likewise, Dade *et al.* (1994) have shown that the thickness of the turbidites decreases away from the source. However, Zeng *et al.* (1991) have noted that the mean grain size decreases downflow often only on the lower fan area (see discussion of submarine fans below). In the upper and middle fan zones, there can be a downflow increase in the main grain size at the channel margins due to overbank splays. The coarse sediment often bypasses the upper and middle fan, where only the fine sediment is deposited by the overbank spreading of the upper part of the current. Rothman *et al.* (1994) have suggested that the distribution of turbidite thicknesses, in the case of minimal erosional truncations and amalgamation, should obey the power law $N(d) \propto d^{-B}$, where N(d) is the number of layers with thicknesses greater than d and $B \cong 1$. Middleton & Neal (1989) have concluded that, if all other variables are held constant, turbidite thickness is directly proportional to the grain size. In an average turbidity current, most of the coarse sediment would be deposited in the time-span of hours, although a complete settling of the fine-grained tail might take a week or so (Kuenen, 1967). Turbidite successions are common in the geological record, which indicates that turbidity currents, despite their highly episodic character, are an important depositional process in deep-water basins.

3.1.2 Debrisflows

Debrisflows are highly concentrated, high-viscosity, sediment mass-flows that have a finite yield strength and thus a plastic-flow behaviour (Johnson, 1970; Hampton, 1972; Boggs, 1987). The sediment and water mixture is fully remoulded by shear, such that any original bedding or stratification are destroyed (Stow *et al.*, 1996). Debrisflows may be turbulent, but the subaqueous turbulent mass-flows are classified as high-density turbidity currents (Lowe, 1982).

46

Rheological Aspects and Types of Debrisflow

In the Coulomb-viscous rheological model (Johnson, 1970), the shear strength (k) has two components, cohesional and frictional, and the constitutive relationship for the flow is:

$$\tau = \underbrace{c + \sigma\prime_i * \tan \phi\prime}_{\text{yield strength } (k)} + \mu * \frac{dU}{dy} \tag{3.3}$$

with the requirement for the flow to occur: $\tau > k$ (at least along the base) where τ is shear stress, c is cohesion, $\sigma\prime_i$ is the effective internal normal stress, $\phi\prime$ is angle of effective internal friction, μ is the coefficient of viscosity and $\frac{dU}{dy}$ is vertical velocity gradient, or the shear-strain rate. On the basis of this rheological model, Nemec & Steel (1984) have classified debrisflows into *cohesional* and *cohesionless*, depending upon which of the strength components predominates. This classification corresponds with the engineering classification of "soil" (natural clastic materials) into analogous two categories (W. Nemec, pers. comm., 1998).

The viscosity coefficient in many debrisflows is roughly constant, independent of the shear-strain rate, which means Bingham plastic rheology (as is typical of cohesive debrisflows; Johnson, 1970; Johnson & Rodine, 1984), but in other debrisflows it may increase or decrease with the increasing shear-strain rate; these are fluids with non-Bingham plastic rheology. Dilative debrisflows, such as a grain-flow, show an increase of the flow's apparent viscosity with the strain rate (Lowe, 1976; Middleton & Southard, 1984). Contractive flows, such as a liquefied sandflow, show a decrease of the apparent viscosity with the strain rate. In short, the "shear-thickening" debrisflows are also "shear-strengthening", whereas the "shear-thinning" flows are "shear-weakening", contrary to one's intuitive expectation based on laboratory shear-box tests.

The velocities of debrisflows vary from the velocity of a fast sediment creep to the velocity of some less robust turbidity currents. Allen (1985) has used the Bingham-plastic model to derive a mathematical model in which the flow velocity, U, is given by:

$$U = \frac{1}{\eta_{ap}} \left(\frac{(\rho - \varrho)g \sin \alpha}{4}(R^2 - r^2) - k(R - r) \right) \tag{3.4}$$

where η_{ap} is the apparent viscosity (Bringham viscosity), ρ is the bulk density of the debrisflow, ϱ is the density of the pore fluid, α is the slope angle, R is the radius of the channel in which the debrisflow moves, r is the radius of the debrisflow's non-shearing "rigid plug", k is the yield strength. Johnson & Rodine (1984) have calculated the critical thickness, z_c, for the initiation or *en masse* stoppage of a debrisflow, assuming Bingham plastic behaviour:

$$z_c = \frac{\left(\frac{c\prime}{\gamma_b \sin \alpha}\right)}{1 - \left(\frac{\tan \phi\prime}{\tan \alpha}\right)} \tag{3.5}$$

where $c\prime$ is effective cohesion, γ_b is the unit buoyant weight, α is the slope angle and $\phi\prime$ is the effective angle of internal friction. The equation is easily verified by letting $FS = 1$ (critical condition) and $\Delta u = 0$ in equation 2.7 and solving for z (calculation for stresses, instead of forces). Equation 3.5 implies, for instance, that a more cohesive debrisflow can attain greater thickness, whereas deeper slopes will cause thinner flows. As noted by Martinsen (1994), the ratio between the critical thickness and the slope angle is important to the internal shear regime of the debrisflow. When the slope angle is at the critical value, the thickness of the flow will be less than the critical thickness, which may mean a shear zone at the base and a "rigid plug" above.

Flow Initiation and Development

Debrisflows are initiated by disintegrative slope failure (see earlier section 2.2). Once initiated, a density flow is able to move on slope gradients possibly not much higher than 0.5° (Stow, 1994), because the subaqueous slopes normally are relatively smooth, fine grained and the flow is supported by buoyancy and dewater slowly, sustaining its mobility (Stow *et al.*, 1996). However, grainflows generally require steep slopes to get started and also relatively steep gradients to move in the basin. The more mobile debrisflows are able to spread over wide areas, to 10s or 100s of kilometres downslope (Stow *et al.*, 1996). Many debrisflows evolve into turbidity currents when accelerating, incorporating water at the top and front, and becoming turbulent (Allen, 1971; Hampton, 1972). The onset of turbulence in a Bingham-plastic flow occurs when (Middleton & Southard, 1984): $\rho U^2/k > 1000$, where ρ is the buoyant density, U is the flow velocity and k is the yield strength. Liquefied flows become turbulent much easier, whereas grainflows seldom reach the necessary speed, as shown by their common sudden "freezing" on relatively steep slopes, of the order of $20 - 25°$ (e.g., foresets of aeolian dunes, giant tidal bars and Gilbert-type deltas). The entrainment of ambient water is stronger when the slope is relatively long and steep (Turner, 1973), and the higher speed causes also the entrainment of bottom silt and sand (Mulder *et al.*, 1997a). Many debrisflows show internal shearing nearly throughout the flow thickness, whereas higher-viscosity debrisflows may be shearing in the basal and possible top zone only, with the flow's core part moving passively as a rigid plug ('frozen' sediment). The lateral edges of a debrisflow are thinner, and the shear stress there is often too small to overcome the material's strength. Levees thus develop, confining the flow and rendering it more mobile. Such debrisflows form elongate, tongue-shaped deposits, whereas lower-viscosity flows spread wider and extensive sheets (Stow *et al.*, 1996).

Debrisflow Deposition

When the downslope pull of gravity no longer exceeds the shear strength of the sediment mass, or the lubricating excess pore pressure has been dissipated, the flow comes to a halt, or 'freezes', by a rapid expansion of the rigid plug to the flow's total thickness (Stow *et al.*, 1996). The deposit of a debrisflow, called "debrite" by some authors, normally retains the original grain-size distribution and sorting (commonly poor), because this flow mechanism involves little grain-size segregation – apart from possible inverse grading. (Normal grading indicates turbulence and implies a turbidity current.) When clast collisions dominate (grainflow), larger particles tend to be pushed upwards, towards the free upper surface of the flow (Bagnold, 1954; Walton, 1983), while the smaller particles may percolate downwards by kinematic sieving (Middleton, 1970; Scott & Bridgwater, 1975). The result is an upward coarsening of particles. Another, cruder type of inverse grading develops when the coarse clasts settle out from the lower, shearing part of the debrisflow and are left behind (Nemec & Postma, 1991). The maximum clast sizes appear to correlate positively with the bed thicknesses in many debris-flow deposits derived from primary sources (Nemec & Steel, 1984).

Debrites are common in slope aprons, or at the foot of a slope, whereas their occurrence in submarine fans is usually limited to the proximal zones.

3.2 Depositional Environments

Depositional environments dominated by turbidity currents and debrisflows include submarine fans, ramps and slope aprons, depending upon whether the sediment source is of a single-point, a multipoint or a linear type. In each of these categories, four further types are recognised according to the predominant grain size: mud-rich, "mixed" mud/sand, sand-rich and gravel-rich systems (Reading & Richards, 1994).

3.2.1 Submarine Fans

Submarine fans are distinctive large-scale depositional features that develop seaward of major point sources of sediment supply, or at the termini of main cross-slope supply routes (Stow *et al.*, 1996). Sedimentological models proposed to explain the development of submarine fans include those given by Normark (1978), Nelson & Nilsen (1974), Mutti & Ricci-Lucchi (1972), Mutti & Ricci-Lucchi (1975), Walker (1978, 1980), Nilsen (1980), Stow (1981), Howell & Normark (1982), Bouma *et al.* (1985), Heller & Dickinson (1985), Mutti & Normark (1987), Shanmugam & Moiola (1988), Posamentier *et al.* (1991), Walker & James (1992), Normark *et al.* (1993) and Ouchi *et al.* (1995).

Submarine fans are formed by turbidity currents, with a minor or major contribution by debrisflows, and are in many respects a seafloor equivalent of alluvial fans. (Some alluvial fans are dominated by turbulent waterflow, channelized or not, whereas others by debrisflows). Sediment accumulation rates are estimated as ranging from less than 1 m per 10000 years to more than 1 m per 1000 years (Stow, 1985). The fan, in its radial profile, can often be divided into the upper, middle, and lower segments, although these are not always distinct (Stow, 1986). The main morphological elements include one or more large feeder channels, or valleys (with depths of 10 to 1000 m; Reading & Richards, 1994), slumps/slides and debrisflow deposits; distributary channels with broad levees; lobes built at the end of the channels (which shift laterally, either gradually or more avulsively); and relatively smooth or current-moulded interchannel and interlobe areas, occupying most of the lower fan in some cases (Stow, 1986). Coarse-grained sediment is distributed radially, through the channels, and deposited as elongate sandbodies or sandlobes. Fine-grained sediment may be distributed laterally by channel overflow into levees and interchannel areas. The levees of deep channels indicate that at least some of the turbidity currents involved must be up to several kilometres wide and several hundred of metres thick (Komar, 1969; Nelson & Kulm, 1973; Stow & Bowen, 1980). The lengths of the channels and the flat abyssal plains indicate that some turbidity currents can travel as far as 4000-5000 km (Curray & Moore, 1971; Chough & Hesse, 1976; Piper *et al.*, 1984). The submarine fan's lenght can vary from about 1 km to more than 2000 km, and the gradients are usually about 2°-5° in the upper fan to less than 1° in the lower fan. Fan areas vary from 1000000 km^2 in large systems to 100 km^2 in moderate systems, to 0.1 km^2 in some gravel-rich systems (Reading & Richards, 1994). Sediment volumes have a similar range.

3.2.2 Submarine Ramps

The term "ramp" refers to broad accumulations of sediment in the lower part of submarine slopes and base-of-slope areas that have multiple point sources in the upslope zone, such as large deltas or well-supplied, channeled shelves. Their downslope lengths range from 50–200 km in mud-rich systems to 1–10 km in gravel-rich systems, with the gradients varying respectively from 0.14–1.4° to 1.15–14.0° (Reading & Richards, 1994). The depositional processes include turbidity currents, debrisflows and slides/slumps, and the sediment thickness may be comparable to those of submarine fans.

3.2.3 Slope Aprons

Slope aprons develop below the edges of relatively narrow shelves, or deep-basin shorelines, and include the continental slope-and-rise zones, the flanks of island arcs, mid-ocean ridges, plateaux and isolated seamounts, the margins of carbonate platforms and the flanks of fjordal basins (Stow *et al.*, 1996). Slope aprons, in contrast to ramps, are fed by essentially continuous linear sources. The basinward extent of a slope apron may range from 10–100 km in mud-rich systems to 1–5 km in gravel-rich systems, with the gradients varying from 2.3–8.5° in muddy to 1.15–26.6° in sandy/gravelly systems (Reading & Richards, 1994). The depositional processes are similar as in the ramps, although lack any pronouced "loci" of sediment accumulation.

Chapter 4

Stratigraphic Modelling of Massflows: Literature Review

This Chapter is a litterature review of the previous attemts to model massflow sedimentation in stratigraphic computer-simulation studies. The review focuses on how a massflow algorithm has been constructed or what weaknesses it bears. The review first considers geometric models and then process-based models, particularly the most recent ones (post–1970). The earlier work, generally less sophisticated, has been summarized in the benchmark textbook by Harbaugh & Bonham-Carter (1970).

The massflow models considered, in accordance with the time-scale resolution adopted in the present study, are mainly numerical models in which massflow sedimentation is averaged over relatively long time periods, but include also models pertaining to individual massflow events.

4.1   Geometric Models

A 2-D computer model called SEDPAK (originally SEDFIL) was developed at the University of South Carolina (see Helland-Hansen *et al.*, 1988; Strobel *et al.*, 1989; Kendall *et al.*, 1991). In the model, the sediment types (mud and sand) are represented by right-angle triangles, in which the area corresponds to the sediment volume and the length of the triangle's base corresponds to the sediment's penetration distance into the basin, measured from the coastline. The basinal cross-section model is an array of columns. Each column is filled with marine sediments up to the sea level, and with alluvial sediments to a surface defined by an "alluvial angle". Each time the area of a sediment column is substracted from the triangles, their heights are correspondingly reduced. This numerical process is repeated until the triangle heights reach the position of the sea level, and then the remaining sediment (area of the sediment triangles), if any, is deposited seaward as a single wedge of offshore sediments. Carbonate sedimentation, tectonic movements, eustasy, compaction (with appropriate curves) and local (Airy) isostatic correction are also included in this program. The modelling of massflows in this program is based exclusively on depositional angles. No sedimentation occurs when the slope becomes steeper than the critical angle of deposition, given as an input parameter. Instead, the sediment bypasses the cross-section seawards until the cross-section flattens to the angle of submarine deposition, which, too, is given as an input parameter.

A computer-simulation program developed at the Shell is discussed by Lawrence *et al.* (1990). This 2-D sedimentation model seems to be quite advanced, although details of the algorithm used are not reported, and is apparently geomentrical, even though Frohlich & Matthews (1991) consider it to be partly process-based. The progradation of the depositional system is based on the principles of accommodation space. The nonmarine cross-section is eroded as a function of elevation and erosion-time constants, based on the work of Pitman & Golovchenko (1983); no deposition is allowed in this part. The alluvial sand/mud ratio is determined using the approaches of Allen (1978) and Bridge & Leeder (1979). In the deltaic or nearshore zone, the sedimentation rate decreases according to an empirically-derived exponential function. Pure sand is deposited above the average storm-wave base, and the sand content fraction further offshore decreases systematically to the maximum storm-wave base. In the program, a near-surface slope failure is a function of the slope angle and the pre-specified values of sediment cohesion, internal friction angle and depth of initial overpressure. The function used for the slope failure seems to follow the theory by Mandl & Crans (1981). The slope instability removes sediment and leads to its deposition further downslope, apparently in accordance with an assumed critical angle (like the angle of marine deposition used in the SEDPAK model above).

A somewhat similar though simpler model, has been described by Ross (1990) and further modified by Ross *et al.* (1995) into a version concerned with the development of slope unconformities and submarine fan deposition . As in the Shell model, the alluvial sand/mud ratio here is determined by following Allen (1978) and Bridge & Leeder (1979). In the marine part of the basinal section, sand is deposited above the wave base and mud further offshore, in deeper water. Unlike in the Shell model, carbonate sedimentation is not considered. Sediment bypass occurs when the slope is overlysteep, or "out of grade", and the development of slope unconformities and onlapping submarine fan/apron systems is "primarily controlled by changing basin physiography". The algorithm thus seems to rely solely on pre-defined angles, as is the case in the SEDPAK program.

The model by Turcotte & Kenyon (1984) disregarded massflow sedimentation as too difficult to model. Likewise, Frohlich & Matthews (1991) avoided massflow modelling to reduce data storage and simplify the graphic output.

Other geometrical models used in stratigraphic simulation programs that have failed to consider massflow sedimentation, even in the cases where subaqueous siliciclastic systems are modelled, include those by Bonham-Carter & Harbaugh (1971), Horowitz (1976), Watts & Thorne (1984), Jervey (1988), Cant (1989), Cant (1991), Watts (1989), Collier *et al.* (1990), Coakley & Watts (1991) and Thorne & Swift (1991).

4.2 Process-Based Models

4.2.1 Dynamic-Slope Models

Lukyanov (1987) has presented a 2-D true depth-dependent (see section 1.4.2) simulation program where the principle of diffusion is used to model the erosion and deposition of one sediment type.

Tectonic movements, eustasy, isostasy and compaction are not considered. The algorithm used for massflow sedimentation is based on three *a priori* defined slope angles (which means that the algorithm is geometrical, although the program as a whole is classified as a dynamic-slope model). In the model, the sediment wedge progrades until a maximum slope angle (α_{max}) is reached at some depth. The 2-D volume of the sediment to fail is determined by making the sediment slope after failure have a linear trend landwards from the point where α_{max} ocurred, using the minimum slope angle (α_{min}). The failing sediment is redeposited further basinwards at another predefined, constant angle. The resulting stratigraphic cross-section is not particularly realistic, because both the slope sediment and the basin-floor sediment have constant angles.

Anderson & Humphrey (1990) have developed a model that combines a 2-D diffusion-based algorithm, a source-weathering algorithm and an angle-of-repose algorithm to handle oversteepened slopes of loose sediment. The latter algorithm causes "bypass" conditions in the sediment transport when the selected angle of repose (40°) is approached. In other words, the algorithm is diffusive, but using a geometric relationship as a trigger for the slope failure. Slopes steeper than the angle of repose are allowed as the initial condition, but the transport rate across them is essentially infinite.

Another 2-D model pertaining to subaqueous siliciclastic sedimentation has been reported by Syvitski *et al.* (1988) and further developed by Syvitski & Daughney (1992) to include nonmarine sedimentation. The model simulates hemipelagic sedimentation, delta-front progradation, proximal slope bypass and downslope sediment transfer by diffusion process. All processes are treated numerically in an explicit manner, with time-step lengths of less than one year. Compaction, istosasy and tectonic movements are not considered. The algorithm for massflow sedimentation, like most of the other corresponding algorithms used in stratigraphic modelling, is based solely upon slope angles. A basin slope steeper than a critical angle specified in the program is bypassed by the transported sediment, and the latter is deposited where the slope is lower than another preselected critical angle. In other words, the algorithm is geometric, relying upon slope angles.

Rivenæs (1992, 1993) in his model (see section 1.4) has not considered massflow sedimentation, but suggested that other transport coefficients (or $K(h)$-curves, see 1.4.2), as proposed by Carson & Kirkby (1972), might alternatively be used. Rivenæs has referred to the Akiyama & Stefan (1986) statement that individual turbidity currents cannot be modelled in terms of a dynamic-slope model (as opposed to a fluid-flow model); what can be considered is rather the net average result of hundreds or thousands of massflow events.

Other dynamic-slope models used in computer stratigraphic simulation programs that have failed to consider massflow sedimentation in the context of siliciclastic subaqueous systems, include algorithms given by Flemings & Jordan (1989), Jordan & Flemings (1991), Christie-Blick (1991), Gaffin & Maasch (1991), Kaufman *et al.* (1991), Sinclair *et al.* (1991), Paola *et al.* (1992), Tipper (1992) and Karner & Driscoll (1997).

4.2.2 Fluid-Flow Models

Bitzer & Harbaugh (1987), have developed a 2-D model named DE-POSIM, which assumes that a fluid flows through the basin, eroding, transporting and depositing sediment in accordance with the flow velocity. The deposition and erosion are controlled by the velocity in that segment of each column of the basinal cross-section that represents water. Basin configuration is considered to be a function of the deposition, erosion, submarine mass movements and compaction. If the slope becomes steeper than a pre-defined critical angle ("maxslope"), a submarine mass failure occurs and causes the slope to decrease. The failure results also in a turbidity current. A second pre-defined critical angle ("minslope") defines as to where the turbidite will be deposited.

Program SEDSIM3, presented by Tetzlaff & Harbaugh (1989), uses 3-D equations for fluid flow derived from the Navier-Stokes formulae. The equations for sediment erosion, transport and deposition are derived from a continuity equation for multi-type of sediment. In order to simulate sediment massflow sedimentation, the initial topography, sediment type, the frequency and volume of massflows, and sediment concentration of the massflows must be specified.

Zeng (1992) has developed a 2-D numerical simulation model of individual turbidity currents and related sedimentation, labelled the

FRS model. The framework of this model consists of three numerical components: a sedimentation model that quantifies particle settling in the current; a rheological model that quantifies the viscosity, density and sediment concentration of the current; and a flow model that characterizes the fluid flow based on equations for the conservation of fluid volume, fluid momentum and sediment volume. The input required to run the simulation for a turbidity current includes the gradient of the bottom slope profile, the sediment grain-size distribution, the fluid volume input, the initial flow-thickness range, the initial velocity range and the flow concentration range.

MOSED3D is a computer model developed by Cao & Lerche (1994) for a 3-D simulation of the erosion, transport and deposition of sediment by a gravity current formed by an instantaneous release of a sediment volume. Two types of input data are required: data for the original basin geometry and the sediment composition of the basal slope, and data for the sediment mass to be released. The successive quanta of sediment released are transported downslope and deposited when their flow energy falls below a critical value. MOSED3D is based on the concept that the gravity current moves along a $3D$ surface according to the balance between gravity and friction. The basic equations governing the flow, as well as the erosion, transport and deposition of the sediment, are derived from Allen (1985).

Mulder et al. (1997b), in a study of the 1979 Nice turbidity current, have given a 2-D numerical method to model a bipartite current that consists of a denser lower part and a less-dense upper part. The equations of Edgers (1981) and Edgers & Karlsrud (1982, 1986) for a visco-plastic flow have been used to determine the speed of the flow's lower part (and of an initial debrisflow). The model assumes that erosion occur as soon as the flow's basal shear stress exceeds the shear strength of the bottom sediment, and the suspension criterion defined by Bagnold (1966) and Middleton (1976) is used to determine the critical velocity of the flow's two parts for the sediment settling to occur. The same criterion is used to find the maximum size of particles that can pass from the lower to the upper part of the flow, in suspension.

4.3 General Remarks

The recent textbook by Slingerland *et al.* (1994) on the simulation of siliciclastic sedimentation in basins is representative of the present status of stratigraphic modelling, as far as the recent advances are concerned. Massflow sedimentation is not even mentioned in this book, and the reason is rather obvious: sediment gravity-flow processes are difficult to model, particularly on a stratigraphic (longer-term, or "macroscopic" *sensu* Tipper,1992) scale. No one of the existing "macroscopic" massflow algorithms is particularly good, perhaps with the exception of the one developed by Lawrence *et al.* (1990). The latter has apparently employed an engineering approach to assess the stability of a sedimentary slope, much like in the recent case, although details of that particular algorithm are not available.

Chapter 5

Description of the Algorithms Used to Model Massflows in DEMOSTRAT

5.1 Massflows

As discussed in Chapter 2, the principal triggering mechanism for sediment gravity flows, leading to transfer of sediment to a deep-water environment, is basin-margin slope failure. An unstable sediment mass fails and evolves into a sediment gravity flow. Turbidity currents and debrisflows are the most important massflow processes of sedimentation in deep-water basins (section 3.1). Among these processes, probably only some of the sustained-type turbidity currents are not directly triggered by failure. With this minor exception, all massflow processes are accounted for by the numerical model given in the present thesis. (The author belives that sustained turbidity currents (see section 3.1.1) can more adequately be modelled, in the DEMOSTRAT framework with the use of an algorithm for hemipelagic sedimentation, which may be developed in the future.) The model used is described in the present chapter.

5.2   The Time-Step Loop

The new time-step loop alorithm developed for the DEMOSTRAT, based on a modification of the one originally employed, is shown in Figure 5.1. The loop is basically the same as the one shown in Figure 1.3, except that special algorithms for slope stability analysis and massflow processes have been adapted. In each time step, the stability of the sedimentary slope is repetively calculated and if part of the sediment appear to be unstable, massflows are triggered and their deposition modelled. The massflow sedimentation algorithm proceeds in a loop until all parts of the basinal slope are stable again. Details of the two special algorithms are explained in the following sections.

5.3   The Instability Algorithm

Since the aim of this study was to model massflow sedimentation, triggered by slope instability, it has been essential to develop an algorithm that numerically assesses the stability of basinal slope in the DEMOSTRAT general framework. This has been accomplished on the theoretical basis of mechanical concepts discussed in Chapter 2.

Theoretically, it is possible to account for the effects of all factors controlling slope stability (as discussed in section 2.2) in terms of the DEMOSTRAT program. However, such an all-encompassing analytical approach would inevitably require specification of parameters that are not practical to be handled on the resolution scale of a dynamic-slope stratigraphic model. For instance, the vertical and horizontal accelerations of every earthquake and the wavelengths/amplitudes and frequency of sea waves at particular time would have to be taken into account as known parameters (see equations 2.8 and 2.9). In short, the inclusion of all the possible factors would render the program inpractical, if not opaque and unreliable.

Therefore, only those processes that are considered to be essential and require reasonable input parameters have been included in the model. The key factors that control subaqueous slope stability are the slope angle and sediment deposition rate (which together determine the sediment mass's shear stress) and the pore pressure related

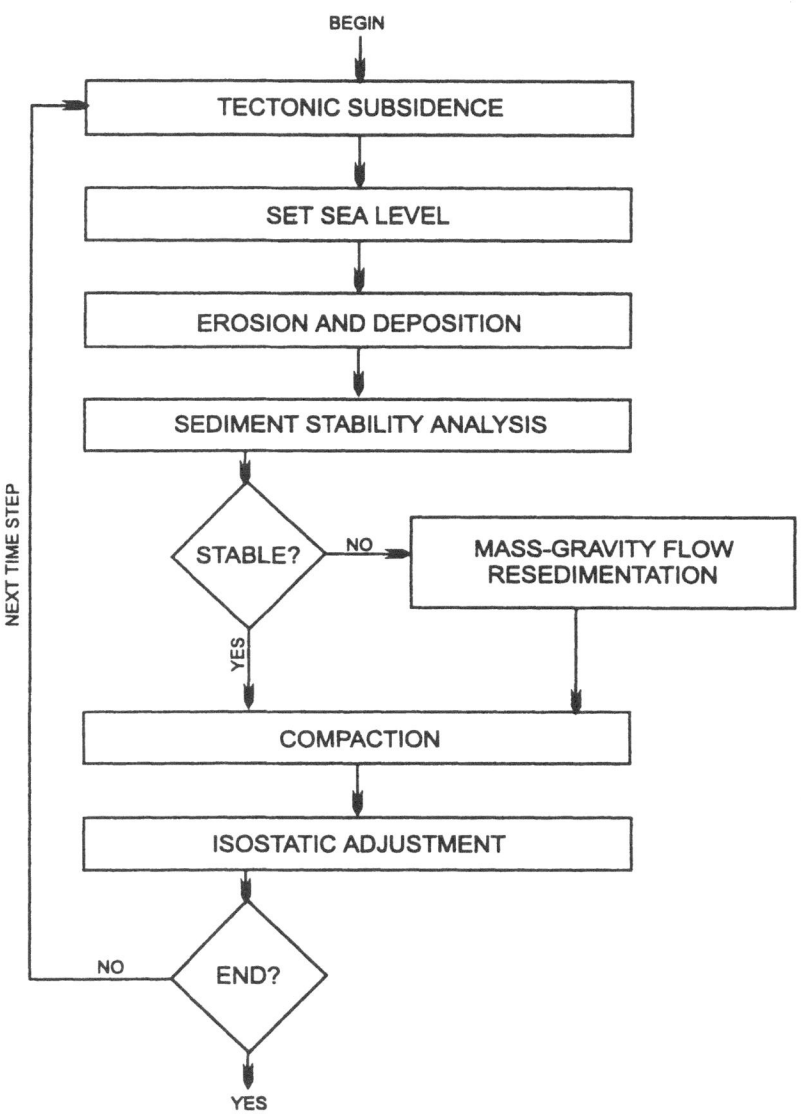

Figure 5.1: The new time-loop algorithm employed in DEMOSTRAT for the purpose of the present study. The processes are modelled sequentially in each time step. All modules, except for erosion and deposition, can be bypassed if unwanted.

to the sediment loading. In addition, a relatively simple statistical algorithm for earthquakes is included. These factors are widely considered to be the most important controls on slope stability, apart from the possible impact of storm waves emphazised by some authors (e.g. Lee & Edwards, 1986). However, one must be aware that large slope failures may generate surficial waves, whose sedimentary record may then lead a stratigrapher to confuse the side effect with the cause.

The instability algorithm to calculate the sedimentary slope conditions is based on the two mechanical theories described in sections 2.1.1 and 2.1.4. The boundary conditions implicit in these theories are thought to be met, or nearly met, by the datasets that DE-MOSTRAT is designed to handle. It is worth noting that the failure plane invoked in the massflow algorithm (described in section 5.4) is not determined in terms of the infinite slope theory, which is here used only to check as to whether the slope sediment is stable or not (see later section 5.3.1). The limiting conditions of a sediment-surface parallell failure plane in the latter theory is thus not crucial.

The instability algorithm, run for each time-step, is as follows: The algorithm first computes FS (see section 2.1.1) for all columns, or the basinal section's unit segments considered, where the margin – or shorline – shed sediment is deposited. The way in which the program computes FS is described in section 5.3.1. If the calculated FS appears to have reached its critical value (see section 5.3.2), the massflow algorithm is initiated. Otherwise, the compaction algorithm is activated (figure 5.1).

5.3.1 Calculation of FS

The factor of safety is calculated for each column basinwards from the shoreline until, possibly, a critical value is found. The calculation proceeds consistently from the left to the rigth, such that if more than one shoreline are present in the basinal cross-section, mass failure will occur in the first unstable slope segment that is encountered in this left-to-right procedure. The latter procedure has no significant implication for the bulk model because the events of resedimentation are considered to be instantaneous. The FS for a given column must be calculated at discrete levels at the column.

These levels are the time-lines crossing the column (Fig. 5.2). In order to evaluate the stability at these levels, it is useful to think of them as potential failure planes. The FS will not be calculated for the last (highest) time-line, for this level is not a potential failure plane. It is also possible to restrict, in the input file, the depth to which the calculation of FS should proceed, simply to reduce the computer running time. This is done through parameter $zmax$: if $zmax$ is taken as a negative value, the instability algorithm will calculate FS all the way down to the first time-line in the column; if $zmax$ is positive, the algorithm will not calculate the FS below $zmax$ (depth in metres). The FS for each time-line is determined by using equation 2.7 (see section 2.1.1) and equation 2.14 (section 2.1.4):

$$FS = \frac{c\prime}{\gamma_b z \sin \alpha \cos \alpha} + \left(1 - \frac{\Delta u}{\gamma_b z \cos^2 \alpha}\right) \frac{\tan \phi\prime}{\tan \alpha}$$

where

$$\Delta u = \gamma_b q t - \gamma_b (\pi \varsigma t)^{-\frac{1}{2}} e^{\frac{-X^2}{4\varsigma t}} \int_0^\infty \xi \tanh \frac{q\xi}{2\varsigma} \cosh \frac{X\xi}{2\varsigma t} e^{-\frac{\xi^2}{4\varsigma t}} d\xi$$

The equation for Δu is solved numerically by a free FORTRAN routine called *gauher* (Press *et al.*, 1992) (see Appendix B), which is implemented in the DEMOSTRAT.

Most of the parameters in the equation for (FS) are specified for "pure" sand and mud, which means that the values for the actual sediment mixtures in the column must be interpolated. Considering a time-line as the potential failure plane, it is clear that it is the sediment directly above the time-line, or the sediment deposited between one time-line and the next (time-step deposition; see Fig. 5.2), that affects the shear stress and shear strength in the sediment at the time-line level. Special care is required by some of the interpolations. In order to calculate the factor of safety for a particular time-line in a column, the following parameters of the equation are to be determined:

> α is the steepest of the two gradients connecting the midpoint of the last time-line in the column with the midpoints of the last time-line in the two adjacent columns (α_1 and α_2 in Fig. 5.2).

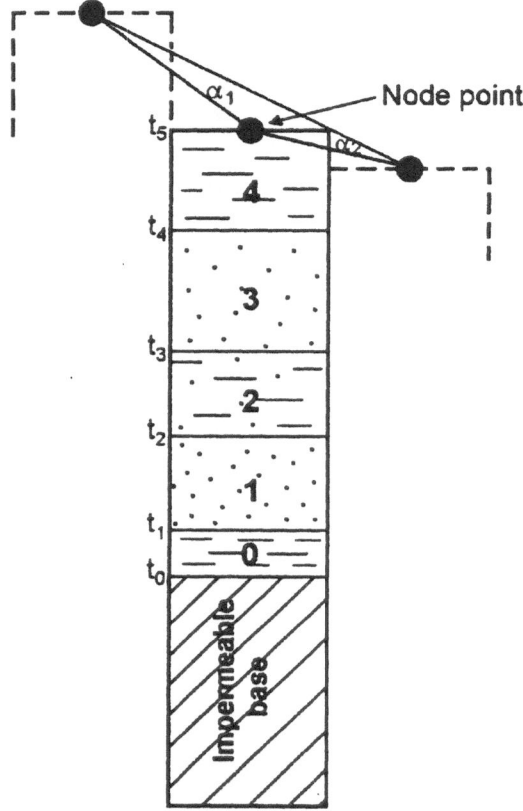

Figure 5.2: Schematic illustration of a basinal column in DEMOSTRAT. In this example, 5 time-lines and 5 time-steps of sediment accumulation (deposits bounded by the time-lines) are shown. The adjacent columns and their node points are indicated by the stippled lines. For further explanation, see text.

$\phi\prime$ is the effective angle of internal friction interpolated linearly from the values given in the input file for pure mud and pure sand on the basis of the (mean) grain-size composition of the sediment deposited in a particular time-step interval directly above the time-line.

γ_b is the mean buoyant unit weight (buoyant weight per volume) of all the deposits above the time-line in the column (this is not true for the γ_b in the excess pore-pressure equation; see below). To calculate γ_b, an estimated value for the average porosity of the sediment in the column above the time-line is used, together with the density of mud (from input file), sand and the surrounding fluid.

z is the total vertical thickness of the sediment above the time-line, computed directly by the program.

$c\prime$ is the effective cohesion, interpolated linearly from the value for pure mud (from the input file) down to zero for an arenitic sand containing $\leq 15\%$ mud, based on sand percentage of the time-step deposits directly above the time-line. The effective cohesion of pure sand is in the program assumed to be zero (see section 2.1.2). This means that possible cohesive effects of capillary forces and cementation (Gray & Leiser, 1982) are not accounted for in the case of pure sand. (However, the effect of early cementation can be accounted for by changing the diffusion coefficients.)

Δu is the excess pore pressure, which has to be treated separately, because the theory that is used to calculate this parameter requires constant-rate sedimentation on an impermeable substratum. The point is to find the principal effect of loading on the sediment column. What is calculated first is the excess pore pressure that would have developed if the column consisted of pure mud (as the equation used is valid for clay deposits). The parameters in the equation are determined as follows: The bottom of the column is the impermeable base, implying that

X is the distance from the bottom of the column to the time-line; q is the mean sedimentation rate for the whole column; γ_b is the buoyant unit weight (buoyant weight per volume) for mud (an estimated value of mud porosity, together with the density of mud (from input file) and the ambient fluid, are used to calculate the buoyant density and further the buoyant unit weight of mud); t is the time from the beginning of the simulation to the last time-step; and ς is the coefficient of consolidation for pure mud (from the input file). This resulting "mud-column excess pore pressure"-value is linearly interpolated, with zero for pure sand. (The trapping of water leading to the buildup of excess pore pressure in the deposit would require extremely high sedimentation rates and is neglected in the program.) The sand percentage that is used in the interpolation to determine the excess pore pressure at a particular time-line level in the column is determined as follows:

- If the percentage of sand in the time-step deposit directly above the time-line (i.e., the time-step deposit of the same consecutive number as the time-line in Fig. 5.2) is equal to or smaller than the percentage of sand in the time-step deposit above (e.g., time-line 2 in figure 5.2), or the higher one if no deposition have occurred, then the percentage of sand in the time-step deposit is used as the sand percentage for the whole column above the given time-line. This is because the excess pore pressure at a given time-line level depens upon the sand percentage in the sediment directly above the time-line and not the sand percentage of the sediment deposited later.

- If the percentage of sand in the time-step deposit directly above the time-line is greater than the percentage of sand in the time-step deposit above (e.g., see time-line 3 in Fig. 5.2), the mean percentage of sand in these two time-step deposits is used as the sand percentage for the whole column above this time-line. (If the time-step deposit directly above is the last time-step deposit in the column, the sand percentage of the particular time-step deposit is used, and if no sediment is deposited in the time-step direcltly above, the sand percentage of the next 'non-zero' time-step deposit is used to calculate the mean.) This is done because in this case, the less permeable layer above wil have a sealing effect.

It is worth noting that no additional flux of fluid from below the time-line, due to permeabilities larger than expected for pure mud, is considered here.

In order to calculate the factor of safety for a particular sediment variety, DEMOSTRAT uses values of three parameters from the input file in addition to the values of the parameters employed in the previous version of the program. These parameters are: cohesion (see section 2.1.2) of pure mud, coefficient of pure-mud consolidation (section 2.1.3) and the angle of internal friction (section 2.1.2) for pure sand and mud. It is to be emphasized that these input parameters can be changed both in time and space during a simulation. One of the advantages of a computer program like DEMOSTRAT is that the effects of parameter changes can readily be seen in the simulation results. All parameters must thus be understood in advance. The parameters have recommended values iserted in the input file for convenience, because the actual values of parameters may not always be available. The recommended values are derived from the engineering literature.

Values of Internal Friction Angles

A range of values $\phi\prime$ for sand and mud (and mixtures of these) can be found in the literature, and some are listed in Table 5.1, which includes also values for the critical angle of repose (see definition below) for sand. The angle of internal friction for a loose sand is approximately equal to the critical angle of internal friction. The values in Table 5.1 are the data basis for the recommended values used in the input file ($\phi_S\prime \approx 30°$ and $\phi_L\prime \leq 20°$).

The angles of repose is well known in the geological and engineering literature. The critical angle of repose (θ_c) is defined as the angle through which a mass of granular material can be rotated before it fails by sliding. The shape, size, sorting, specific gravity and surface characteristics of the sediment particles have the greatest effect on the critical angle of repose (Van Burkalow, 1945; Miller & Byrne, 1966; Carrigy, 1970). Departure from the spherical form and increasing roughness of the surface of the particles increase the critical angle of repose. For the same sediment, the latter is smaller in water than in the air when the sediment is wet or submerged and larger when it is air-dry (Van Burkalow, 1945; Miller & Byrne, 1966; Carrigy, 1970). The angle of rest (θ_{re}) is defined as the inclination of the slope after the mass-failure process has ceased.

Values of Cohesion

Pure, uncemented sand and gravel have essentially no cohesive strength (Middleton & Wilcock, 1994). DEMOSTRAT operates with two "pure" sediment types (sand and mud) that are allowed to mix, with a linear interpolation of the equations for the two types. Cohesion of mud in DEMOSTRAT is specified in the input file.

The cohesive strength of muddy sediment (measured in kPa) is several orders of magnitude smaller than that of mudstones (measured in MPa) (Middleton & Wilcock, 1994). Busch & Keller (1983) have noted that the effective cohesion shows no consistent relationship with sediment composition, texture or the depth below the seafloor. Further, the magnitude of the cohesion in bulk is a function of the nature and number of bonded interparticle contacts per unit volume (Gillott, 1968). Not many empirical values of the effective cohesion of sediments are found in the literature. Busch & Keller (1983) give $c\prime$ values ranging from 1.83 kPa for a relatively coarse, partly bioclastic sediment from the Peru-Chile continental mid-slope, to 18.76 kPa for the overconsolidated mud lens in the upper slope there. Johnson (1984) has reported an effective cohesion of approximately 23 kPa in loose mud.

Values less than 10 kPa is recommended for effective cohesion of pure mud in DEMOSTRAT, although it may not necessarily be a good approximation for other sediment varieties.

Table 5.1: A list of ϕ and θ_c values derived from the literature.

Sediment	Angle	Degrees	Reference
London Clay	ϕ	20	Skempton & DeLory (1957)
clay	ϕ	20-28	Booth *et al.* (1985)
mud	ϕ	20-35	Morgenstern (1967)
silty clay	ϕ	21-26	Almagor & Wiseman (1978)
silty clay	ϕ	24-30	Booth *et al.* (1985)
fine grained sediment	ϕ	24-28	Hampton *et al.* (1978)
fine grained sediment	ϕ	22-28	Sangrey & Marks (1981)
mud, average	ϕ	28	Booth *et al.* (1985)
clayey silt	ϕ	27-33	Booth *et al.* (1985)
silt to clayey silt	ϕ	22.8-29.8	Lo *et al.* (1997)
clayey silt	ϕ	27-29	Jibson (1992)
silt	ϕ	30-35	Booth *et al.* (1985)
silt/sand	ϕ	28-34	Morgenstern (1967)
dune sand in water	θ_c	34.5	Carrigy (1970)
dune sand in water	θ_c	35.0	Carrigy (1970)
sand	θ_c	35	Carrigy (1970)
sand	θ_c	35	Carrigy (1970)
sand	ϕ	35.5-36.4	Lade (1993)
Hostun sand	ϕ	34.4	Schanz & Vermeer (1996)
Hostun sand	ϕ	34.8	Schanz & Vermeer (1996)
Fort Peck sand, sub-rounded	ϕ	33	Kenney (1984)
Sahara sand, sub-rounded	ϕ	34	Kenney (1984)
Ottawa sand, rounded	ϕ	27	Kenney (1984)
Crushed sandstone, angular	ϕ	37	Kenney (1984)
Crushed slate, angular	ϕ	37	Kenney (1984)
Ham River sand, loose	ϕ	33.5	Bishop (1973)
Ham River sand, dense	ϕ	38.3	Bishop (1973)
Ham River sand, very dense	ϕ	44.0	Bishop (1973)
qrushed quarts in water	θ_c	37.1	Carrigy (1970)

Values of Consolidation Coefficient

In equation 2.14, it is the coefficient of consolidation that determines the consolidation state of the deposit, because that parametre controls how fast a sediment can be consolidated. The larger the value of the coefficient, the faster the consolidation occurs and the less over-pressure can develop in the deposit. This means that the coefficient of consolidation in DEMOSTRAT is of great importance in determining the stability factor, which is not a desirable relationship because of the uncertainties associated with the former coefficient. However, the coefficient functions here as an order-of-magnitude variable and its available estimates are believed to be satisfactory for the present purpose. Anyway, this assumption will be confirmed or modified by further parameter calibration in the future.

Some values of the coefficient of consolidation for fine-grained sediments found in the engineering literature are listed in Table 5.2. These empirical values are the basis for the recommended value of this parameter inserted in the input-file of DEMOSTRAT. The recommended value of the coefficient for pure mud is in the range of $1 * 10^{-9}$ to $1 * 10^{-8} \frac{m^2}{sec}$.

5.3.2 When is FS Critical?

The probability of slope failure is not a simple, linear function of FS. On the basis of case studies, Athanasiou-Grivas (1978) has demonstrated that the probability of a failure is low at values of $FS > 1.3$ and that failure is a virtual certainty at $FS < 0.9$ (see Fig. 5.3). In DEMOSTRAT, it is assumed that every failure is or leads to a disintegrative failure (see section 3). This is because the dynamic-slope models, like DEMOSTRAT, have relatively large time-steps, which makes it certain that more than one massflow will originate from an instability during a single time-step. On the basis of this, it is recommended that the massflows in DEMOSTRAT are initiated when $FS < 1.2$ (value in the input-file). (It is probable that a massflow is triggered when the shearing force is up to 0.2 times smaller than the resisting force (Fig. 5.3); wave-induced cyclic stress and intrastratal gas generation, which are is not accounted for by the equation for FS, may further increase this latter factor. Further, it is assumed

Table 5.2: Values of the coefficient of consolidation derived from the literature.

Sediment	Value (in $\frac{m^2}{sec}$)	Reference
clay	$3.9 * 10^{-7}$	Gibson (1958)
clay	$4 * 10^{-8}$	Lambe & Whitman (1979)
clay	$1 * 10^{-7}$	Busch & Keller (1983)
clay	$4 * 10^{-8}$	Craig (1987)
lean clays	$1 * 10^{-6}$	Terzaghi (1955)
colloidal clays	$1 * 10^{-10}$	Terzaghi (1955)
clay	$1 * 10^{-9} - 1 * 10^{-8}$	Morgenstern (1967)
montmorillonite	$.06 * 10^{-8} - 0.3 * 10^{-8}$	Mitchell (1993)
illite	$0.3 * 10^{-8} - 2.4 * 10^{-8}$	Mitchell (1993)
kaolinite	$12 * 10^{-8} - 90 * 10^{-8}$	Mitchell (1993)
clay	$1 * 10^{-8} - 378 * 10^{-8}$	Kondner & Vendrell Jr. (1964)
silty clay	$3.2 * 10^{-9}$	Booth *et al.* (1985)
silty clay	$1.6 * 10^{-8}$	Booth *et al.* (1985)
silty clay	$1 * 10^{-8} - 1 * 10^{-7}$	Morgenstern (1967)
mud	$2.5 * 10^{-8}$	Busch & Keller (1983)
mud	$3.2 * 10^{-8} - 6 * 10^{-8}$	Richards & Hamilton (1967)
fine grained	$1 * 10^{-7}$	Richards & Hamilton (1967)
fine grained	$1.6 * 10^{-7} - 7.4 * 10^{-7}$	Buchan *et al.* (1967)
fine grained	$3.2 * 10^{-9}$	Booth *et al.* (1985)
silty clay	$1.8 * 10^{-7} - 2.3 * 10^{-8}$	Almagor & Wiseman (1978)
clayey silt	$8.2 * 10^{-9}$	Been & Sills (1981)
clay and/or silt,	$3.2 * 10^{-7} - 3.2 * 10^{-10}$	Booth *et al.* (1985)
silt	$3 * 10^{-5}$	Lambe & Whitman (1979)
silt	$1 * 10^{-7} - 1 * 10^{-6}$	Morgenstern (1967)
silty	$3.5 * 10^{-7} - 8.2 * 10^{-7}$	Roberts & Cramp (1996)
coarse silt	$1 * 10^{-6}$	Morgenstern (1967)
clayey silt	$4.5 * 10^{-8} - 4.7 * 10^{-8}$	Hampton *et al.* (1978)
silt to clayey silt	$9.8 * 10^{-7} - 2.8 * 10^{-8}$	Lo *et al.* (1997)

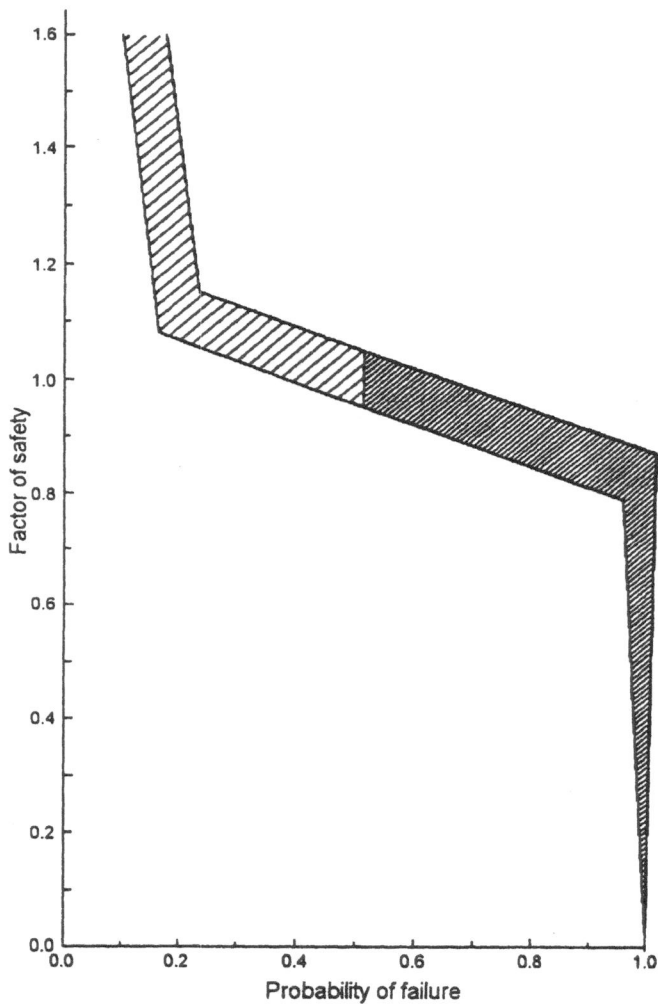

Figure 5.3: Plot of the probability of failure versus the factor of safety (FS) based on case studies (from Booth *et al.*, 1985, ;modified from Athanasiou-Grivas (1978)). Note the contrasting patterns in swath about 50 % probability and the very low probability of failure for $F \geq 1.3$.

that earthquakes cause slope failures when $FS < 2.0$ (value included in the input file).

An earthquake, in contrast to other stress-inducing prosesses, can in an instant of time produce stress that exceeds this strength and thus lead to a mass failure. As mentioned earlier, equation 2.8 is not applicable in dynamic-slope programs, and requires knowledge of parameters that are too specific. However, a simple procedure for a failure triggered by earthquake is included in the program. A random generator of natural numbers is used to yield values from 1 to 100; any yielded value higher than a pre-specified critical value will mean an earthquake to occur. The function draws one value for each time-step. The probability (specified as a natural number, in percent) of a relatively powerful[1] earthquake to occur within one time-step must be specified in the input file, and can possibly vary in time and/or space, but the recommended value is otherwise zero. For example, if a 50-% probability for a powerful earthquake to occur during a particular time-step, any value from 51 to 100, if drawn by the random generator during this time-step, will cause an earthquake to occur. The earthquake will cause failure, and hence a massflow if FS is less than 2.0 and greater or equal to 1.2 (a failure will anyway happen if $FS < 1.2$). Since muddy sediments have cohesive strengths, earthquakes are considered to be more important in triggering massflows when the slope is mud dominated.

5.4   The Massflow Algorithm

The DEMOSTRAT, as a dynamic-slope numerical model, required that the massflow algorithm be adjusted to the model's mathematical design and the information involved. In other words, the algorithm had to be successfully adopted to the operational modules of the program. However, the information processed in DEMOSTRAT ap-

---

[1]Powerful enough to trigger a slope failure. For example, Inouchi *et al.* (1996) have found that earthquakes with accelerations stronger than 44 gal triggered turbidity currents on a fine-grained sedimentary slope of Lake Biwa, Japan. Papatheodorou & Ferentinos (1997) note that sediment failures in some area determined by the epicenteral distance, in the Gulf of Corinth, is expected to occur when the seismic shock exceeds a magnitude of 6.0 in the Richter scale.

peared to be insufficient, so that the resolution could not be as high as in fluid-flow models, which are generally closer to the reality (section 4.2.2). Even if the information were sufficient, the computation time would be far too excessive. In short, it is not possible to model the individual massflows, but rather the net effect of their series corresponding to a time-step considered. The more time-steps are used in a simulation, the higher the resolution, but also the longer the computation time needed. This means that the modeller must have a good basic understanding of massflows (hence the background given in Chapter 3) in order to interpret and predict the results of a simulation with confidence.

When the instability algorithm (section 5.3) causes the program to initiate the massflow algorithm, it has been determined that the solution (the slope profile, or time-line) of the previous time-step is unstable and that massflows will now be triggered.

The massflow algorithm proceeds as follows: First, the zone of failure is defined (in order to minimize further computation time). There can be, in general, several zones where failures can be triggered, as is the case if sediment is supplied to both margins, and these zones must be specified by the user (as either a function of sea depth or a function of slope angles). It is recommended here that the whole zone from the shoreline to the first point where the basin-slope profile of the previous time-step changes from being convex to being concave upwards (i.e., where $\frac{\partial^2 h}{\partial x^2} = 0$; see Fig. 5.5) is defined as a zone of failure. If the failure is for some reason expected to occur outside such a zone, it must necessarily be defined differently (see section 5.4.1).

Second, for each column in the failure zone (from left to right), the FS is calculated for each column, in the same way as in the instability algorithm. If no critical value is found to have been reached in this zone, the algorithm proceeds to the other potential instability zones (the basinal section may include two slopes (margins) but a single slope or a series of slopes (fault blocks) can be used, if more desirable). If a critical FS is found, the algorithm will remove sediment from the zone and deposit it further basinwards, outside this zone. Section 5.4.1 explains as to how the gravitational erosion, transport and deposition take place. A single episode of gravitational erosion is called an *instability episode* (IEP), and the associated episode of sediment redeposition is called a *depositional episode* (DEP). After the IEP and DEP, the FS is again calculated in all columns in each instability zone, alternatingly, until a new IEP–DEP occurs. This process continues until all columns (in all failure zones, if several) have uncritical FS values. Several IEP and associated DEP can occur within any given time-step. The massflow algorithm is summarized in Figure 5.4. When the process is done, the program will initiate the compaction algorithm (see Fig. 5.1).

5.4.1 Erosion, Bypass and Deposition

Erosion

The zones of potential failure, as defined in the previous section, must be specified by the user directly in the input file. The zones can be defined on basis of both water depth and slope gradient. Consequently, their widths can be customized, so to speak, if there are reasons to assume that the failures will involve a wider or a narrower zone than the recommended one.

The gravitational erosion (mass failure) in a given zone is executed by making *pseudo time-steps*, with the diffusion equations (section 1.4.2) run with a small t and large diffusion-coefficient values. The idea of letting the time be short and the diffusion coefficient large is to reflect the instantaneous nature of a mass failure. The time, in the pseudo time-step, used in the equations is adjusted, by a program loop, so that the critical FS found by the instablility algorithm will have a value of 2 (i.e., not too close to instability, because mass failure renders the remaining sediment more stable rather than reaching the limit of stability, i.e. $FS = 1$).

The diffusion coefficients will also be tuned outside the failure zone(s), so that both transport and deposition occur within the same pseudo time-step (see further below). The amount of the diffusivity-coefficient increase is specified in the input file.

The amount of sediment (2-D volume) to be removed by the mass failure and flow is specified by the time-line (slope profile) of the pseudo time-step (see Fig. 5.5). The average sand percentage for this sediment mass is determined by the algorithm.

It is utterly important to understand that each instability episode

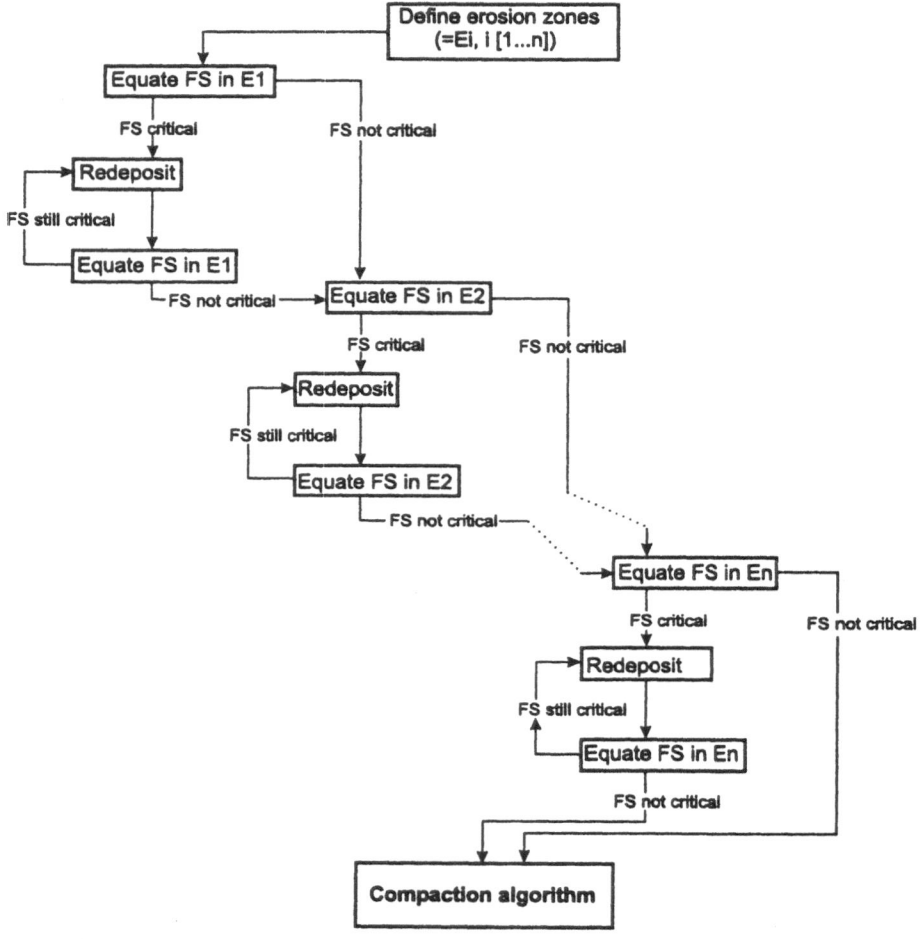

Figure 5.4: Graphical summary of the massflow algorithm used.

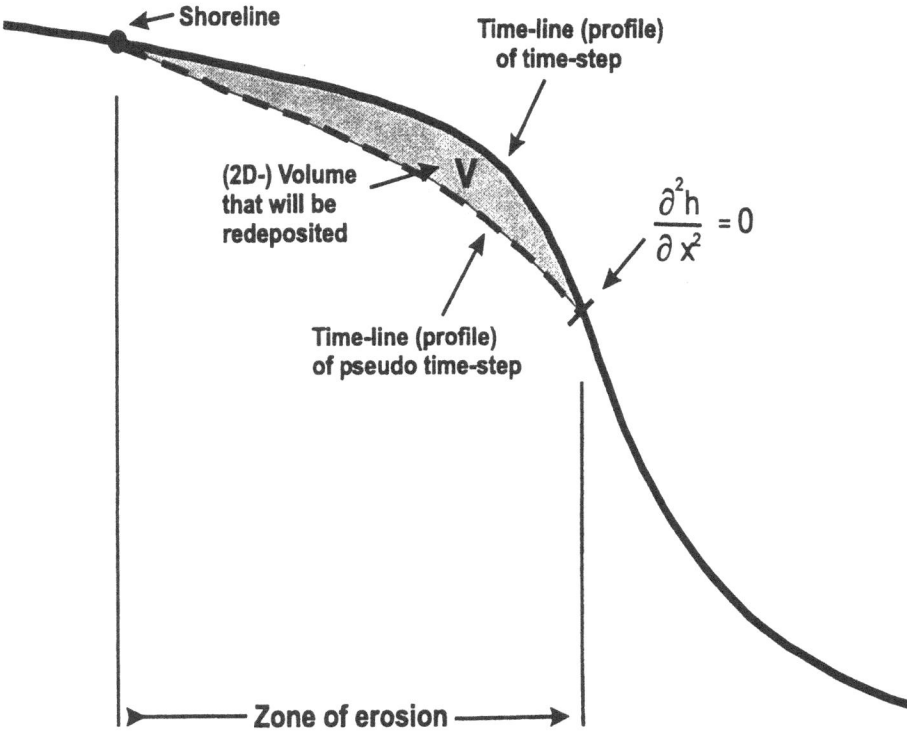

Figure 5.5: Erosion occurs by increasing the diffusion coefficients for the columns (not shown here) of the zone of erosion in a pseudo time-step. Note the 2-D volume of sediment to be removed by the mass failure and flow in the pseudo time-step.

will in reality consist of several massflows (see Fig. 5.6). The downslope transfer of the sediment mass is here called a massflow event
(MEV). The successive mass failures involved in a given IEP will normally be retrogradational which means that the successive massflows
will tend to be more sandy and generated progressively closer to the
sand-rich shoreline. Every MEV is necessarily followed by a depositional event (DEV) further downslope, where the sediments removed
by the failure is deposited. This means that one DEP in the program
will in nature consist of several DEVs. Since the MEVs in each IEP
normally be more sandy with time, each DEV in a DEP will be more
retrogradational (see below). This concept is summarized in Figure
5.6.

Deposition

The average sand content of the 2-D volume of the failing sediment
(Fig. 5.5) together with basinal slope determines as to where and
how this sediment will be distributed in the basin in a DEP. That
is, of course, because volume, concentration, grain size, and speed
for each massflow are not known. In the present model, the average
deposition resulting from all the DEVs in a given DEP is considered,
which is the optimal approach suitable for DEMOSTRAT. Intuitively,
one might argue that there must be a point in the section where
the cummulative deposition from the massflows in a DEP begins (see
PSEPA in Fig. 5.6). The deposits upstream of this point are thought
to be volumetrically insignificant on a basin scale and most of them
will anyway be "in transit", only termporalrily deposited. This is in
accordance with the common assumption that the spatial thickness
distribution of a sedimentary body composed of massflow deposits
will, in the longitudinal section, resemble Gaussian (or asymmetri-
cal Gaussian) distribution (W. Nemec, pers. comm., 1997). This
assumption would imply that 99.7 % of the 2-D volume of the body
is within the distance of ± 3 standard deviations from the mean
thickness of the body. Further, the point defining the transition
from bypass to deposition, for otherwise equal sedimentary bodies
with different mean grain sizes, are thought to happen at different
basinal gradients. This point can thus be related to, and determined by, the mean grain size of the sediments involved in an IEP.

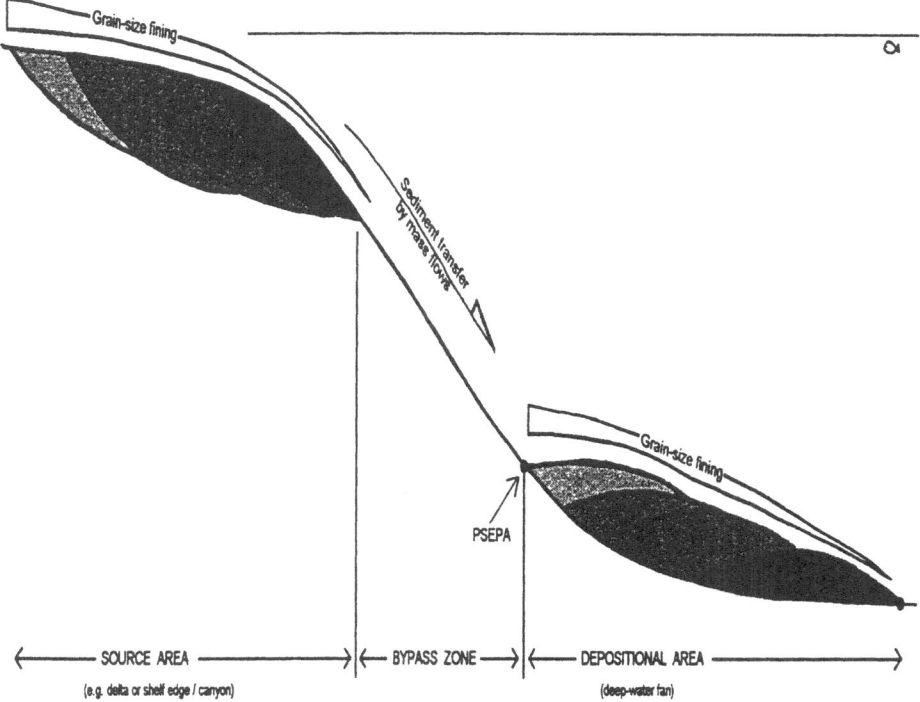

Figure 5.6: An instability episode (IEP) and the associated depositional episode (DEP), represented by four massflow events (MEVs) and corresponding depositional events (DEVs). PSEPA = point of significant episodic aggradation. See text for further explanation.

The point is here called the *point of significant episodic aggradation* (PSEPA) (see Figs. 5.6 and 5.8) and the basinal gradient at that point is called the *critical sedimentation gradient* (CSG). It is likely, of course, that some of the massflows in the real world will be deposited on steeper slope. However, these deposits will be liable to further resedimentation. In short, these mass-flow "extrema" is not particularly significant to the model.

As mentioned in section 3.1, the principal massflows delivering sediment to a deep-water basin are turbidity currents and debris-flows, of which the former are much more important. Based on section 3.2 and the references therein, the CSG in the program is assumed to be 6° for pure sand and 0.5° for pure mud. Linear interpolation is used for sand–mud mixtures. The PSEPA, for sand and mud can be selected in the input file. The linear interpolation means that mud-rich flows will travel farther than sand-rich which is consistent with the empirical observations reviewd by Reading & Richards (1994) and the deep-water efficiency concept of Mutti (1985).

Even if a DEP in the program corresponds to several DEVs in nature, the very possibility of having several DEPs within one time-step renders the mathematical model as close to the reality world as is possible in terms of a dynamic-slope framework, because the sediment grain size, for determination of the CSG, will be averaged over smaller volumes than if only one DEP per time-step were allowed.

A summary of how the gravitational erosion and deposition take place in the real world, compared to the present model is shown in Figure 5.7, for both the shelf source and the basin-floor area of deposition. In the real world (Fig. 5.7 A), the time/volume relationship for an overall deposition (EGD) has a positive gradient, whereas the relationship for an overall erosion (EGR) has a negative gradient. The erosion in the source area is due to massflow events (MEVs), each with a corresponding depositional event (DEV) in the depositional area. Furthermore, hemipelagic sedimentation in periods without massflow deposition, renders the time/volume gradient in the depositional area to be slightly positive. In the model (Fig. 5.7 B), the erosion in all MEVs and the deposition in all DEVs occurs simultaneously. No hemipelagic sedimentation is considered to occur, whereby the time/volume gradient in the depositional area is zero during periods with no massflows.

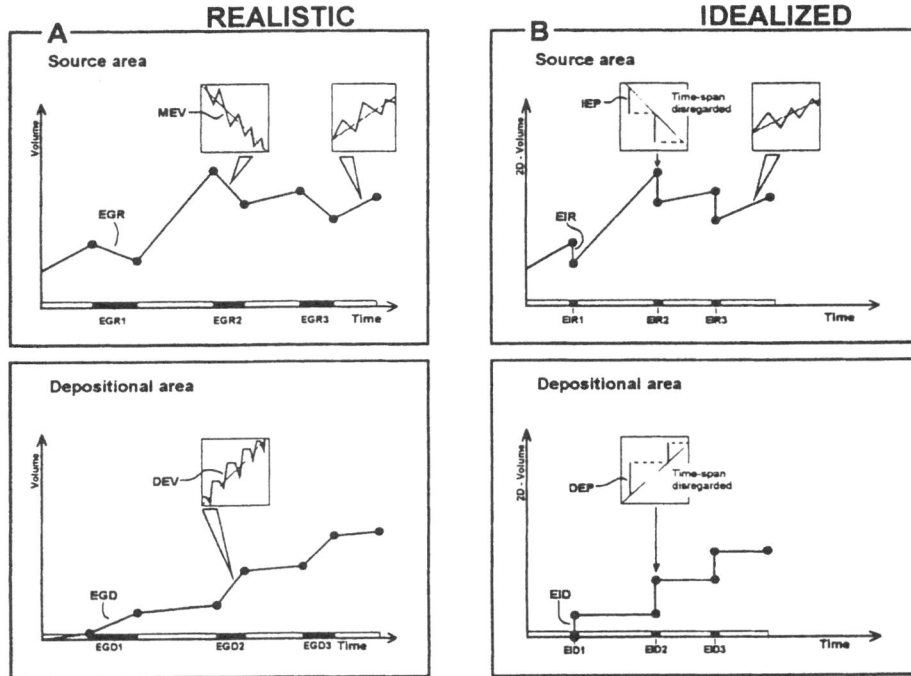

Figure 5.7: Comparison of erosion and deposition between the real world and the model. MEV = mass-flow event; EGR = episode of gradual removal; DEV = depositional event; EGD = episode of gradual deposition; IEP = instability episode; EIR = episode of instantaneous removal; DEP = depositional episode; EID = episode of instantaneous deposition. See text for further explanation. The figure was inspired by the Kolmogorov (1951) model of sedimentation.

It is worth noting in Figure 5.7 that the episodes of gradual removal (EGRs) are associated with massflows and basin-floor aggradation, but not all EGRs must necessarily be associated with the generation of massflows (e.g., forced regression without massflow sedimentation). If the latter was the case in the EGRs in Figure 5.7, episodes EIR 1-EIR 3 would be identical to episodes EGR 1-EGR 3. The time between the initiation of a massflow and its deposition is disregarded in Figure 5.7. The time-span in the IEPs and DEPs is disregarded because their duration is neglibly short in nature.

The deposition basinwards from the PSEPA in the pseudo time-step is modelled using intermediate values for the diffusion coefficients above the previous time-line level, and lower values below that level. This makes the deposition occur on top of the latter surface. The diffusion coefficients used for the deposition has to be specified in the input file, which contains no recommended values. This gives the modeller a free choise. The massflow diffusion coefficients are inserted in the same way as it is normally done in DEMOSTRAT (see Rivenæs, 1993).

Transport

The massflow transport of sediment takes place between the basinward boundary of the erosion zone and the PSEPA (see Fig. 5.8). The transport in the pseudo time-step is provided by using large values for the diffusion coefficients above the time-line level of the time-step, and low values below that level. The transport then occurs without slope erosion. The values for the diffusion coefficients need to be specified in the input file (as usual for DEMOSTRAT).

The diffusion coefficients landward from the inner, landward boundary of the erosion zone should be very low, so that no sediment transport occurs in this zone. No recommended values is contained in the input file.

When the algorithm has completed the erosion, transport and deposition in all of the basin, the time-line for this last time-step is replaced by a time-line that follows the previous pseudo time-line in each zone. Where two pseudo time-lines meet, as a result of deposits merging in the basin centre, the replacing time-line will follow the top defined by the merged pseudo time-lines.

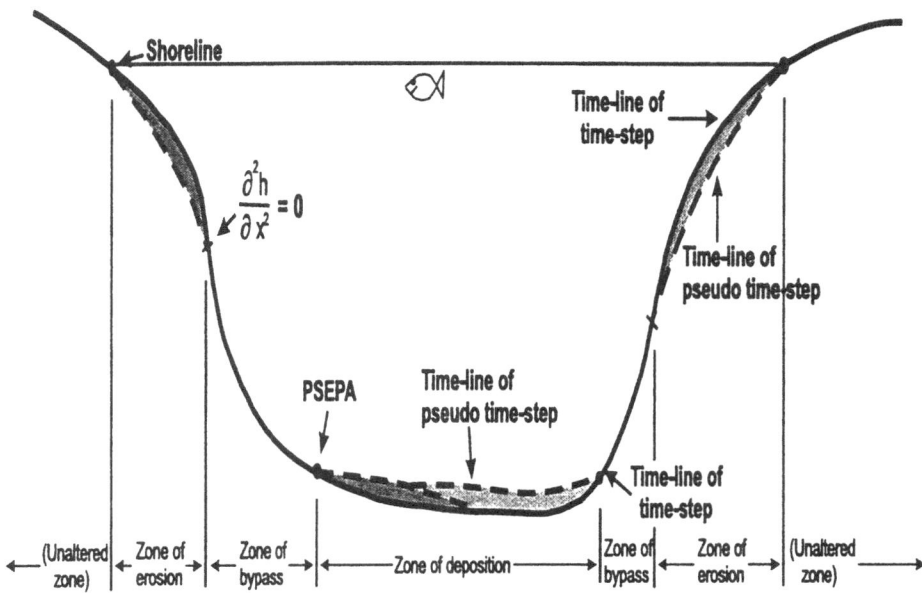

Figure 5.8: Schematic illustration of how the massflow algorithm functions in DEMOSTRAT. Only one time-line is shown. The algorithm resediments the eroded 2-D volume (coloured parts directly below the shorelines), beginning the resedimentation at the PSEPA (point of significant episodic aggradation), which is determined by the mean grain size of the 2-D volume. Only one IEP and one DEP are shown on each basin side. The massflow deposits merge in the centre. This will not always be the case, depending upon the diffusion-coefficients values used for the depositional zones. See text for further explanation.

A summary of the massflow algorithm is given in Figure 5.8, which shows two basin-margin failure zones with associated zones of bypass and deposition. In a basinal setting like this, the two deposition zones will merge in the centre.

5.5 A Simple Alternative for the Instability Algorithm

An alternative, simpler method to model massflow sedimentation is also implemented in DEMOSTRAT. The simple method is equal to the one described, except for the triggering mechanism involved, which is not based on slope-stability analyses. Instead, massflows are triggered if the gradient of a time-line becomes greater than a pre-defined critical value. In other words, the triggering mechanism is not dependent upon sediment type, thickness, cohesion, pore pressure and angles of internal friction, but on the critical gradient alone.

This method resembles many of the methods reviewed in Chapter 4, although the erosion, transport and deposition of massflows here follows a process-based model, contrary to the pre-existing massflow models, which are mainly geometric (see details in Chapter 4). The simple method was implemented here merely to check certain aspects of the program regarding gravitational erosion, transport and deposition. The method itself is not as good as the one described above. Nevertheless, both methods are now included which allows comparisons to be made and to assess the importance of cohesion, pore pressure and internal friction (that all depends on sediment type) to the slope sediment stability.

Which of the two methods to employ, has to be specified in the input file.

Chapter 6

Synthetic Simulations

In order to illustrate as to how the computer program works and can possibly be used in future studies, a range of example simulations are reviewed in the present chapter. The input parameter values used in these simulations are synthetic, not meant to reflect any particular natural case although the geometries of the simulated basin-fill successions may be representative of some fan deltas. The discussion will focus on slope instability and massflow sedimentation.

The reader may notice small discrepancies between the algorithms described in the book's sections 5.3 and 5.4 and the algorithms implemented here in the DEMOSTRAT program, but these are merely adjustments made in an early stage of the implementation. The discrepancies and their effects on the simulations are as follows:

- Parametres $c\prime$ (effective cohesion), $\phi\prime$ (effective internal friction angle) and ς (consolidation coefficient) are taken to be constant, unchanging in either time or space.

- The boundary between the bypass zone and the depositional zone (PSEPA) is assumed to be independent of the sediment type, in contrary to what is described in section 5.4.1. For a massflow, the deposition zone begins where the slope angle decreases to a value specified in the input file, which means that the "efficiency" of the massflows in a DEP is not fully modelled in the present simulations. As a consequence, CSG will be identical for all sediment mixtures and the massflow deposition will thus be retrogradational, backlapping the basin-floor

87

slope, because the local gradient of the latter will decrease as a result of the sediment deposition. This will not be the case, of course, if the slope clinoforms prograde over the base-of-slope depositional wedge.

- The time parameter in the pseudo-time steps here is not tuned up to make the factor of safety equal to 2.0, as described in section 5.4.1, but the diffusion coefficients are tuned up, such that slope sediment stability in the pseudo-time steps is achieved. In short, this will have no effect on the reality of a simulation.

- The earthquake algorithm is not employed, as its relevance is limited to tectonically active areas. For simplicity, the basin is assumed to be free of seismic disturbances.

In the present simulations, the sediment is supplied to only one margin of the basin, such that the resulting style of massflow sedimentation is relatively easy to observe. The supply rates are taken to be 1 m²/yr for sand and 1 m²/yr for mud (porosity ignored). The three zones distinguished in section 5.4.1 (failure zone, bypass zone and depositional zone) are here defined on the basis of sediment surface inclination. The time step in all simulations is 1000 years. The basin's cross-section has a total length of 20 km and is divided into 400 columns (the whole basin length is not shown in the print-out figures). The initial cross-sectional morphology of the basin can readily be seen in Figure 6.1 (top of the grey-shaded area).

Simulations 1 to 4 are meant to demonstrate as to how the basic process of massflow sedimentation is modelled, and to show the difference between the results of massflow triggering by the slope stability method and the critical angle method (described in sections 5.3 and 5.5, respectively). The simulations involve no changes in the relative sea level and employ constant rates of sediment supply, in order to keep the picture of massflow sedimentation as simple as possible. The effects of isostasy and compaction are included in simulation 4.

Simulations 5 to 7 demonstrate that the present model of massflow sedimentation can be used to analyze various stratigraphical problems and verify concepts of a basin-fill history. These three simulations correspond to simulations 1-3, respectively, but involve the effects of compaction, isostasy and eustatic sea-level changes. The latter are assumed to have a sinusoidal pattern.

6.1   Simulation 1

The result of simulation 1 (Figure 6.1) shows the basin-fill situation after 50,000 years of a constant sediment supply, with a stable sea level and no isostatic or compaction effects. The original version of DEMOSTRAT program has been used in this simulation, with no massflow sedimentation being modelled. The sedimentary wedge has prograded into the deep-water basin, with the slope angle becoming increasingly steeper (although eventually converging due to the nature of the diffusion equations used). After 50000 years of sediment accumulation, the steepest slope is about 17°, and the slope angle will continue to increase so long as the water depth increases. If the simulation continued under these conditions, the slope would thus become even steeper, which is an unrealistic result, especially with respect to the large sediment input involved ($2\frac{m^2}{yr}$).

6.2   Simulation 2

Simulation 2 was fully analogous to simulation 1, but involved mass-flow sedimentation – with the slope failures triggered by the simple method based on slope angle, as described in section 5.5. The sedimentary wedge had initially prograded and steepened similarly as in the previous simulation, until a critical angle of 15° was reached and massflows were triggered. The simulation log-file indicated the first series of massflows to have occurred in time-step 38, around 4 km away from the basin margin, at an elevation of about 180 m[1] (see Figure 6.2). The massflows bypassed part of the slope and commenced their deposition where the slope angle was about 1.4° (ca. 8 km away from the basin margin). The effect of erosion is readily recognisable to the right of the first slope-failure escarpment, and the

---

[1]Elevation in this chapter is invariably relative to a datum level set by the program and hence the same as the 'elevation' axes in the simulated cross-sections. In other words, the elevation here is not a distance from the sea level.

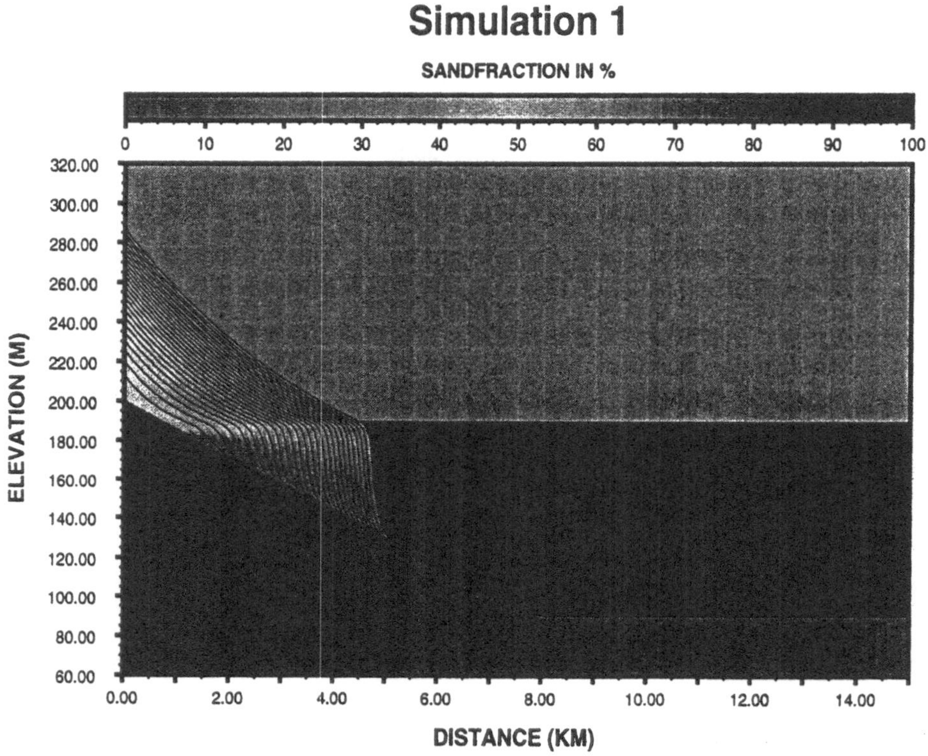

Figure 6.1: Simulation 1: basin-fill cross-section after 50,000 yrs of steady sediment supply. The sea level was assumed to be stable; no effects of isostacy and compaction were included; and no massflow sedimentation was involved. The basin's initial topography is shown by the top of the shaded bottom part. Every second time-step is indicated with a black solid line.

corresponding deposits have been spread beginning at ca. 7 km distance from the basin margin. As the sedimentary wedge continued to prograde, two more IEPs and DEPs occurred during time-steps 44 and 49. The general pattern of the massflow deposits is slope back-lap, because the deposition here is assumed to begin at a critical slope angle that is independent of the sediment type. This simplification, though numerically convenient, should be eliminated in a later stage of the program's development and implementation.

In order to trigger massflows, a critical angle as low as 15° has been selected, although other computer models use considerably higher critical angles. For example, Anderson & Humphrey (1990) used a critical angle as high as ca. 40°. If a similar critical value, or half of it, were used in simulation 2, no massflows would have occurred at all. Even if the simulation were run over a much longer time, the sediment slope would be unlikely to ever probably reach an angle of 40° anyway. This would not be a realistic scenario, considering the relatively high sediment input. However, to decrease the critical angle might not be a good solution either, because it is by no means certain that massflows in reality would necessarily be triggered every time the slope gradient reaches the reduced critical angle (which is here lower than both the angle of internal friction and the angle of repose for each sediment type).

6.3   Simulation 3

The result of simulation 3 (Figure 6.3) portrays a depositional system analogous to that in simulation 2, except that the massflow triggering here has been controlled by the sediment-stability criterion described in section 5.3. The values of the input parameters assumed in the stability analysis are as follows: the angle of internal friction for pure sand is 30° and for pure mud is 20°, the coefficient of consolidation is $0.1 * 10^{-8}\frac{m^2}{s}$ and the cohesive strength is 1 kPa. Massflows are triggered when FS becomes less than 1.2.

In simulation 3, eight episodes of massflow sedimentation occurred (during time-steps 26, 29, 33, 37, 41, 43, 46 and 49). In Figure 6.3, the resulting failure scars are recognisable at distances of 2 to 4 km away from the basin margin, at elevation above 180 m,

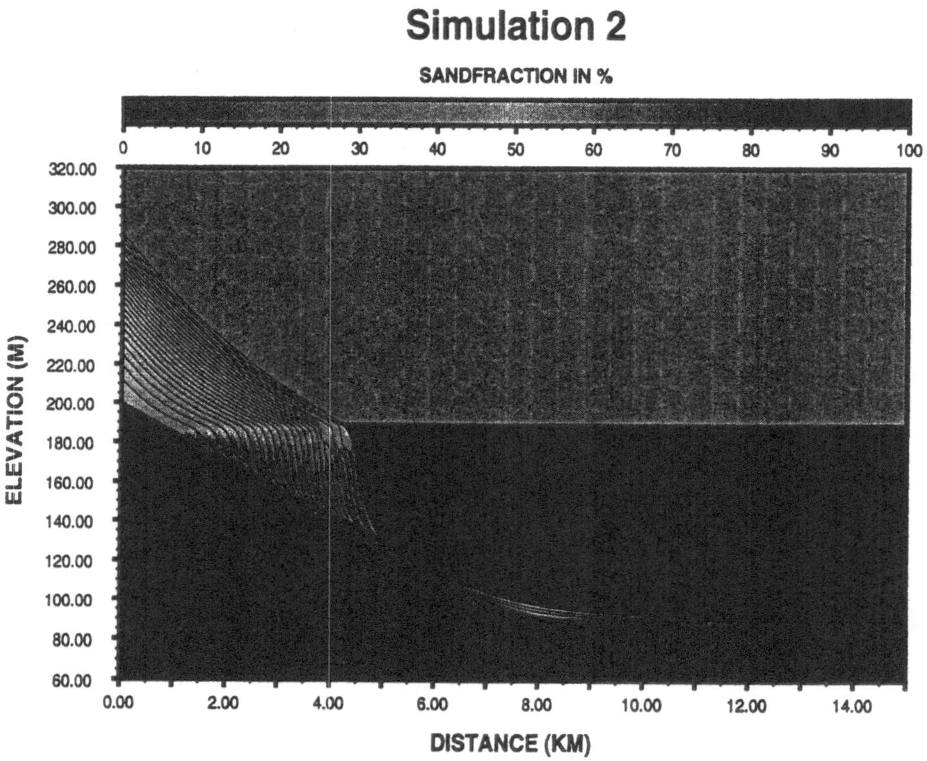

Simulation 2

SANDFRACTION IN %

Figure 6.2: The result of simulation 2 (for details, see text).

and the corresponding back-lapping deposits occur basinwards from a distance of 8 to 6 km. The simulation log-file printout shows the surface angle in each column and the interpolated values of excess pore pressure, cohesion and FS, among other parameters. The values of these parameters show that the sediment stability is not merely related to the slope angle (obviously, since the stability is governed by equations 2.7 and 2.14). For example, Table 6.1 shows some of the log-file values for column 83 at time-step 43, for time-lines 41 and 42. Here, the sediment surface angle ("angle (deg)") is, of course, the same for both time-lines. There is a notable change in the sand fraction above these time-lines (as indicated, for instance, by the difference in the cohesion value) and the sediment thickness ("rzthick") above each time-line in the column, which results in greater pore pressure at the level of time-line 41 than at the level of time-line 42 in the column. This causes FS to decrease below 1.2 at the level of time-line 41 (slope instability), but exceed the critical value at the level of time-line 42 (slope stability).

The simulation demonstrates that it is practically impossible for the angle-based simple method to define a reliable single critical value at which massflows would be triggered. Massflows can be triggered at a whole range of slope angles, depending upon the sediment type, effective cohesion, excess pore pressure and thickness (above the potential failure plane), as specified by equations 2.7 and 2.14.

6.4 Simulation 4

The difference between simulation 4 and simulation 3 is the inclusion of the effects of compaction and isostasy in the present case. These two factors contribute to the steepening of the sedimentary slope, and hence the simulation result (Fig. 6.4) shows a more frequent recurrence of slope instability, with episodes of massflow sedimentation during time-steps 26, 28, 31, 33, 35, 37, 39, 41, 42, 44, 46, 47, 48 and 49. The sedimentation time represented by this simulation is the same as in the previous examples (50,000 yrs).

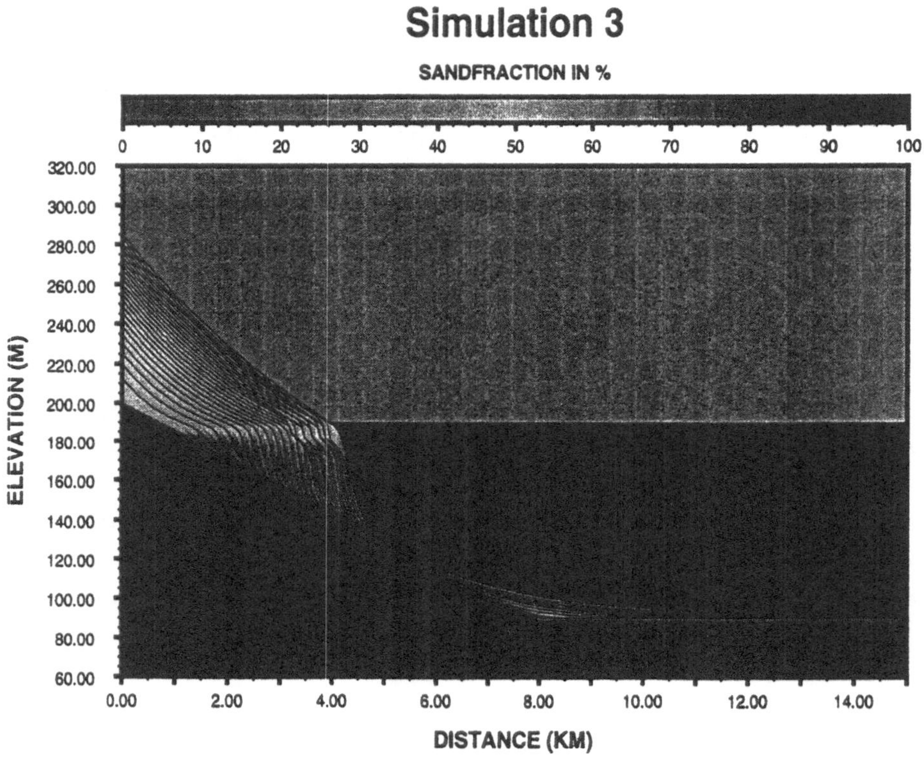

Figure 6.3: The result of simulation 3 after 50,000 yrs. Note the similarity to the result of simulation 2 in Fig 6.2. The massflow trigging process here is controlled by the slope stability analysis described in section 5.3.

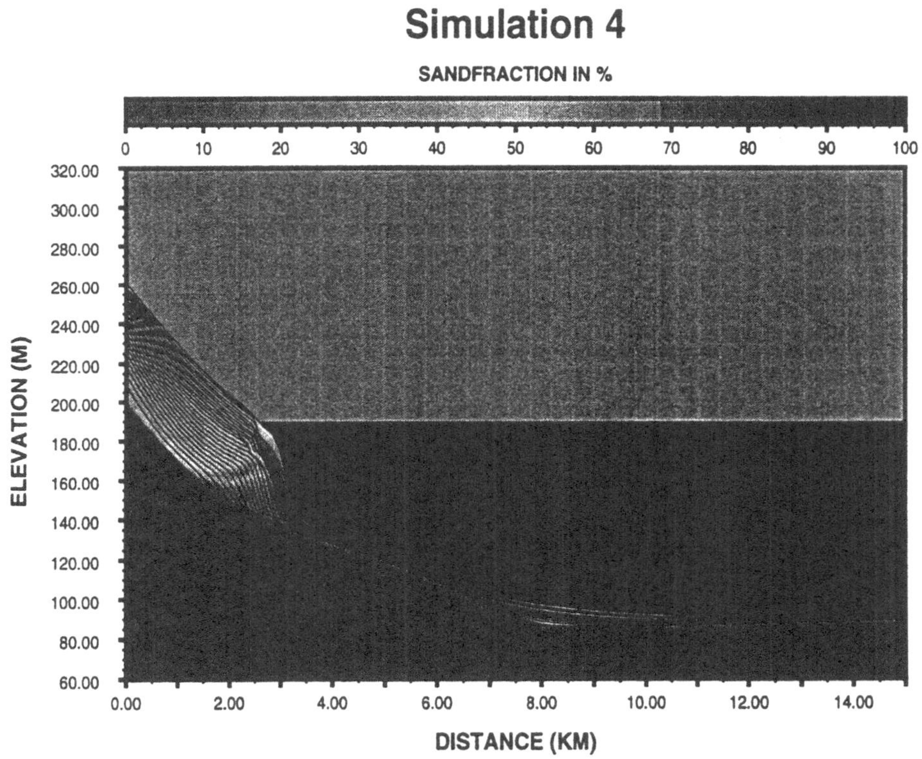

Figure 6.4: The result of simulation 4, including the effects of of compaction and isostasy (for details, see text).

Table 6.1: Simulation 3 – example of log-file printout (for column 83, time-step 43, time-lines 41 and 42).

Time-step: 43		
Column: 83	:	**Time-line: 42**
FS	:	1.260
angle (deg)	:	11.10
delta u	:	16939.79
cohesion	:	909.56
rzthick	:	9.98
Time-step: 43		
Column: 83	:	**Time-line: 41**
FS	:	1.162
angle (deg)	:	11.10
delta u	:	21722.95
cohesion	:	980.08
rzthick	:	12.28

6.5 Comment on Simulations 1-4

The simulations demonstrate that the computer model of massflow sedimentation does work and gives reasonable results. They show further that the use of a single critical angle as the trigger criterion for slope instability, as employed in some other models, is not only simplistic, but also quite problematic. If a critical angle close to the sediment's angle of repose, or angle of internal friction, is assumed, little or no slope instability will occur. In reality, slope steepening is not the sole factor responsible for massflow triggering. If a much lower critical angle is assumed, the recurrence of slope failures will be unrealistically high, and there is no way in which a "realistic" (optimal) critical angle might possibly be selected – because the sediment properties will likely vary with time. Therefore, the criterion for slope instability, to be reasonably realistic, must necessarily include other factors (as described in section 2.1).

Consequently, one must conclude that the existing computer programs for massflow sedimentation adopted in geometrical and dynamic-slope models (e.g. Lukyanov, 1987; Helland-Hansen *et al.*, 1988;

Syvitski *et al.*, 1988; Ross, 1990; Ross *et al.*, 1995), as reviewed in sections 4.1 and 4.2.1, are rather unrealistic. They not only model massflow sedimentation in straight lines (by applying a constant angle to define the sediment's top surface in the source area as well as in the zones of erosion and deposition), but render the timing of the slope failures highly arbitrary and questionable.

6.6   Simulation 5

The result of simulation 5 (Figure 6.5) shows, again, a basin-fill case similar to that in simulation 1, but including the effect of eustatic sea-level changes (as shown in Figure 6.6). In addition, the effects of compaction and isostasy have been included. Like in simulation 1, however, no massflow sedimentation is involved in the present case; the basin slope merely accumulates sediment and progrades. It is worth noting that the surface angle of the basin-margin subaerial (alluvial) deposits seems to be much steeper than it is in reality, because of the cross-section's exaggerated vertical scale (Figure 6.5). In fact, the highest angle in this part of the cross-section is about 1.7°, which is quite steep, but not unrealistic for a delta slope. The high-frequency "disturbances" shown by the time-lines (Figure 6.5) are a result of compaction, not slope failure.

6.7   Simulation 6

This simulation (Figure 6.7) differs from the previous one in including the effects of slope failure and massflow sedimentation. The time of sediment supply represented is 100,000 yrs. The simplistic criterion of a single critical slope angle (section refsimple) is used for the massflow triggering, but the critical angle has here been adjusted (see discussion in section 6.5), such that the number of the massflow sedimentation episodes would be the same as in simulation 7, where slope stability analysis has been used as the failure criterion. The critical angle used is 6.5°, and the episodes of massflow sedimentation have occurred during time-steps 8, 10, 12-19, 21, 52, 54, 55, 57-59, 60-72, 74, 76, 93, 95, 96 and 98-100 (which means 38 times).

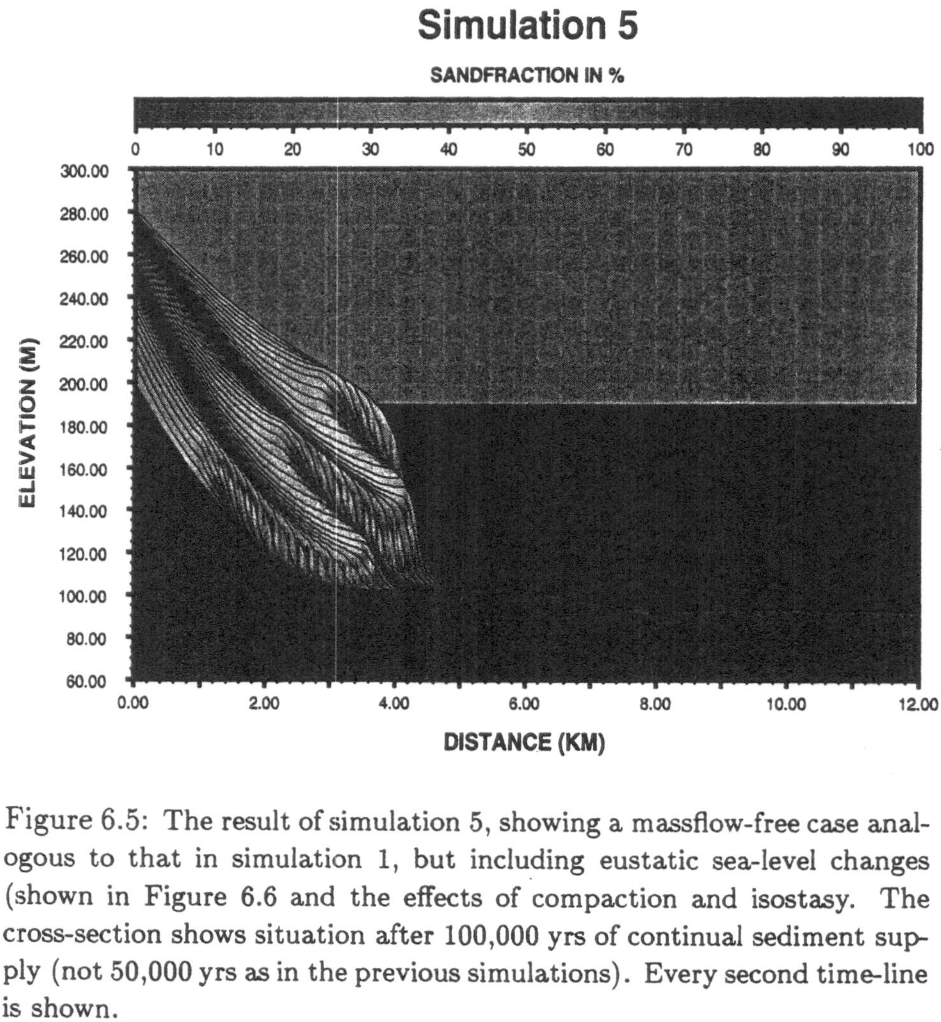

Figure 6.5: The result of simulation 5, showing a massflow-free case analogous to that in simulation 1, but including eustatic sea-level changes (shown in Figure 6.6 and the effects of compaction and isostasy. The cross-section shows situation after 100,000 yrs of continual sediment supply (not 50,000 yrs as in the previous simulations). Every second time-line is shown.

EUSTASY VS TIME

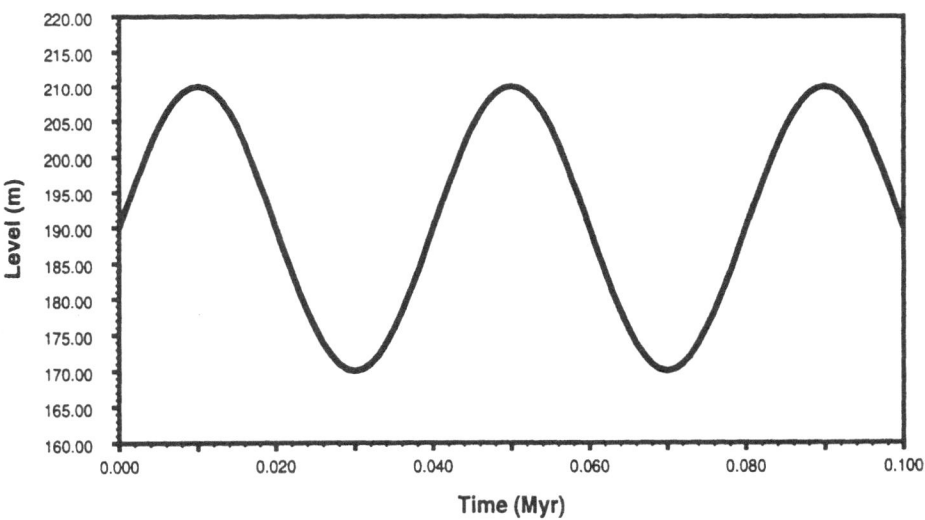

Figure 6.6: The curve of eustatic sea-level changes used in simulations 5 to 7. The amplitude is 40 m and the period is 40,000 yrs. Sea-level oscillations of this order may be glacioeustatic (see Plint *et al.*, 1992, fig.1).

The massflow sedimentation occurred in three time periods (Figure 6.9): from 9,000 to 21,000 yrs, from 51,000 to 76,000 yrs, and from 92,000 to 100,000 yrs. The massflow deposition starts at a distance of about 4.5 km from the basin margin.

6.8   Simulation 7

The only difference between this simulation (Figure 6.8) and the previous one is the use of slope stability analysis as the failure criterion (same as in simulation 3). The values of the input parameters for the stability analysis are: the angle of internal friction is 30° for pure sand and 25° for pure mud, the coefficient of consolidation is $0.1 * 10^{-8} \frac{m^2}{s}$ and the cohesive strength is 1 kPa. Massflows have been triggered when the FS decreased below 1.2, with the episodes of massflow sedimentation during time-steps 9, 43-60, 81-85 and 87-100 (which means 38 times). The massflow sedimentation thus occurred mainly in two time periods (figure 6.10): from 42,000 to 60,000 yrs and from 80,000 to 100,000 yrs, with a minor episode in the 9th time-step. The slope-failure escarpments are readily visible at the transition from the more sandy to the more muddy part of the subaqueous part of the basinal cross-section (Figure 6.8), and the effects of massflow deposition can be seen beginning at a distance of about 5 km from the basin margin.

The result and its comparison with that of simulation 6 show clearly that the criterion of a single critical slope angle, even when adjusted to yield a "realistic" number of failure episodes, cannot be expected to give the same result as that based on the more realistic criterion of multiparametric slope-stability analysis.

6.9   Comments to Simulations 6 and 7

It is worth noting the differences in the timing of the episodes of massflow sedimentation in the two simulations (Figs. 6.6, 6.9 and 6.10). In simulation 6, massflow sedimentation first occurs during the period from 9,000 to 21,000 yrs, during a normal regression

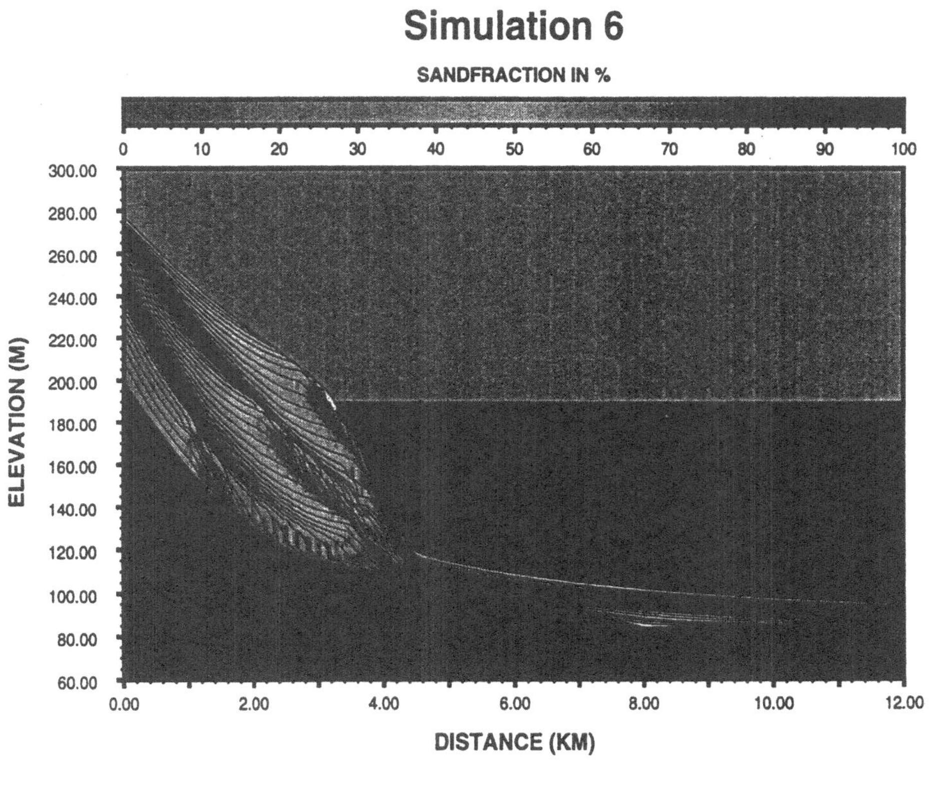

Figure 6.7: The result of simulation 6, representing 100,000 yrs of continual sediment supply. The case is similar as in simulation 5, but includes massflow sedimentation – as in simulation 2.

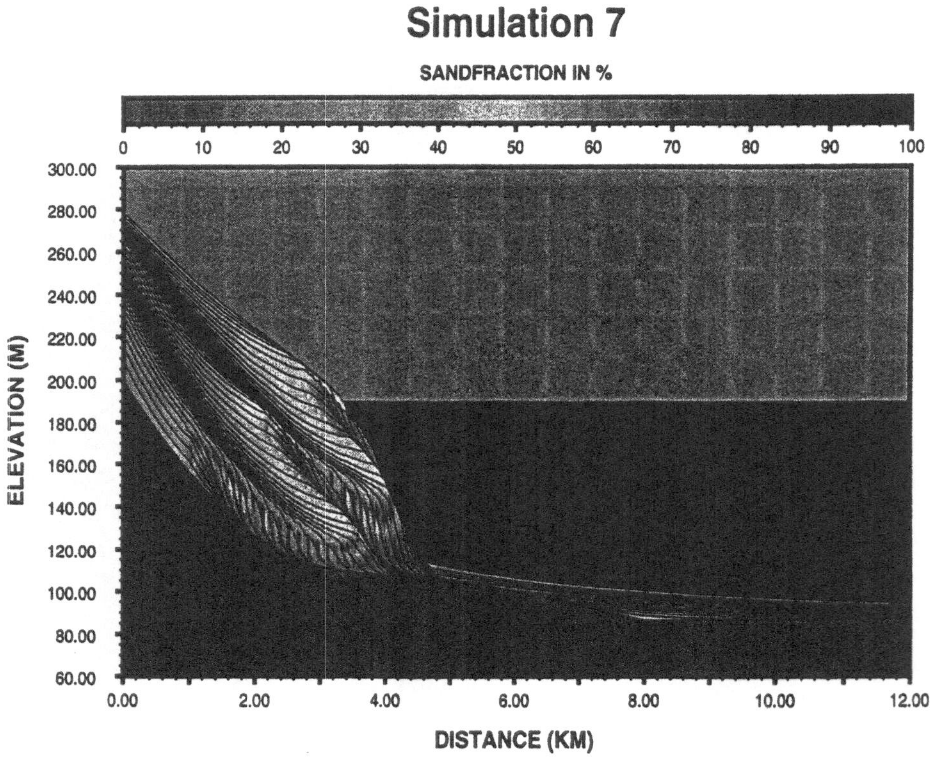

Simulation 7

SANDFRACTION IN %

Figure 6.8: The result of simulation 7, showing a case similar to that in simulation 6, but involving the criterion of slope stability analysis – as used also in simulation 3. The cross-section shows basinal situation after 100,000 yrs of continual sediment supply.

(HST[2]), with only two slope-instability events, and forced regression (FRST). In the period from 22,000 to 50,000 yrs, no massflow sedimentation took place during the last phase of the forced regression and the following normal regression (LWST), transgression (TST) and renewed normal regression (HST). Massflow processes then occurred in the period from 51,000 to 76,000 yrs, with two DEPs during the transgression (TST) and further slope failures during the following forced regression (FRST) and normal regression (LWST). No massflows took place during the subsequent transgression and normal regression (HST) from 77,000 to 91,000 yrs, but new slope failures occurred during the forced regression (FRST) from 92,000 to 100,000 yrs.

In simulation 7, one DEP occurred during the first normal regression (HST) (time-step 9), but no massflow sedimentation took place during the subsequent forced regression (FRST) and normal regression (LWST) and the first part of the following transgression (TST), from 10,000 to 41,000 yrs. Massflow sedimentation occurred in the last part of this transgression and during the following normal regression (HST) and early forced regression (FRST) from 42,000 to 60,000 yrs. No slope instability occurred in the period from 61,000 to 79,000 yrs, during the last part of the forced regression and the following normal regression (LWST) and early transgression (TST). Massflow sedimentation was then renewed in the late phase of this transgression and the following normal regression (HST) and forced regression (FRST) from 80,000 to 100,000 yrs.

Despite several similarities, simulations 6 and 7 clearly show some interesting differences in the timing of massflow sedimentation. Notably, no massflow processes occurred in simulation 7 during the normal regressions at the times of low relative sea level (LWST), as might be unexpected (see section 7.2). In contrast, massflow sedimentation occurred during the transgressions and both normal and forced regressions in simulation 6, although was sparse in the TSTs (i.e., during transgressions) and HSTs (i.e., during a normal regression at the time of high relative sea level). Furthermore, the massflow sedimentation in simulation 6 was more scattered in time

---

[2]The four-fold division of stratigraphic sequence, as discussed by Helland-Hansen & Martinsen (1997), is used in the present case.

than in simulation 7.

It is also interesting to note that the episodes of massflow sedimentation in simulation 7 that lack time-equivalents in simulation 6 are all related to the late phases of TSTs and in HSTs. These episodes in simulation 7 are likely to have been related to the sand-poor slope conditions of these phases and triggered by excess pore pressures, although other factors might play a role, too (see discussion in section 2.2). The lack of massflow sedimentation in the LWSTs can be explained by the short time span (less than 8000 yrs) of these systems tracts in the present case (see Figs 6.7, 6.8 and 6.6).

Although simulation 7 may be considered to be more realistic than simulation 6, it would be unwise to draw general conclusions about the timing of massflow sedimentation in a sequence stratigraphic framework. In the light of these simulations, no particular systems tract seems to be preferentially prone to massflow processes. More simulations, with a wider range of realistic input data, have to be performed before more specific conclusions can possibly be drawn.

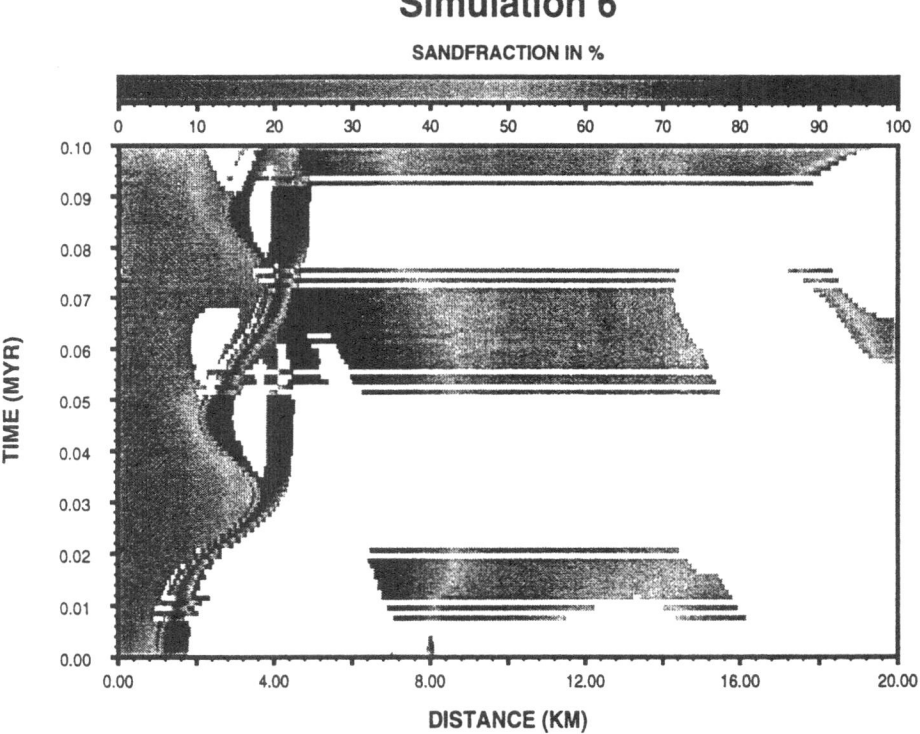

Figure 6.9: Time-stratigraphic diagram for simulation 6, showing a cross-section of the basin-fill sedimentation with time. Based on the principle of Wheeler (1958, 1964).

105

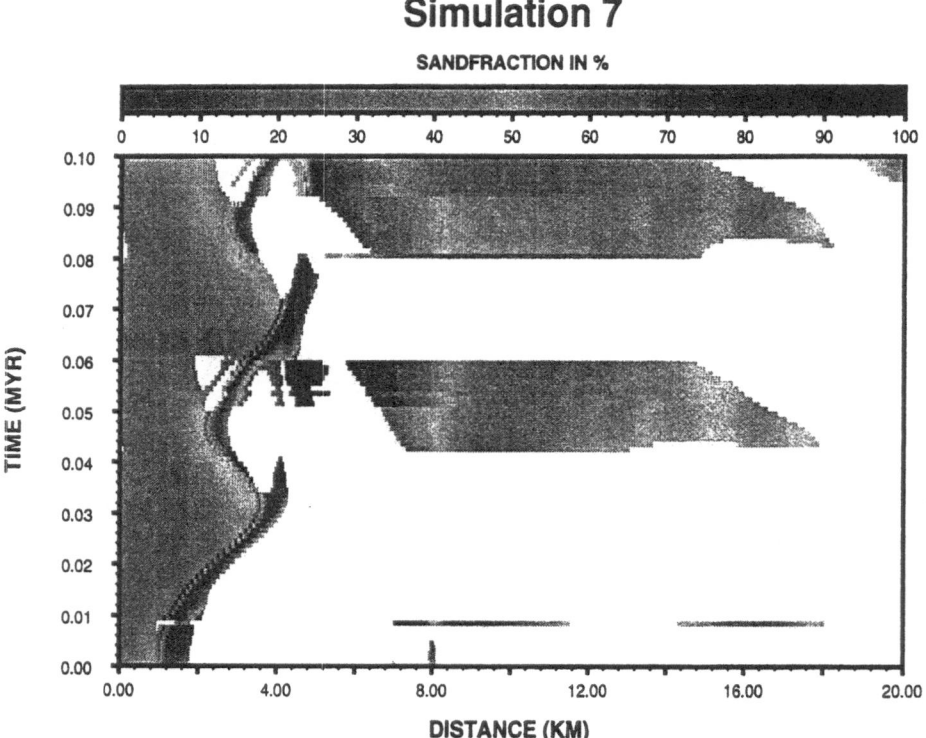

Figure 6.10: Time-stratigraphic diagram for simulation 7, showing a cross-section of the basin-fill sedimentation with time. Based on the principle of Wheeler (1958, 1964).

106

Chapter 7

Discussion

7.1 Choice of Model

The stratigraphic simulation model presented in this book is meant to operate within the numerical framework of DEMOSTRAT, therefore its time-resolution scale has been selected to match that of the latter dynamic-slope program (section 4.2.1). For example, a fluid-flow approach and a computer model with a time-resolution scale comparable to that of single massflows would be quite unsuitable in the present case. On the other hand, a fluid-flow approach is not particularly suitable for stratigraphic computer models because of the huge amount of input data and numerical calculations required. The time resolution in the present model has been made as detailed as is practical possible in a dynamic-slope approach, by introducing a program loop in the massflow algorithm that makes it possible to have several IEPs and DEPs within any single time-step.

A geometrical massflow algorithm might be an alternative approach, as used in dynamic-slope models by some other authors (e.g., Lukyanov, 1987; Syvitski *et al.*, 1988). However, the combination of different approaches often renders a simulation program rather opaque, more difficult to follow and control. It has thus been a challenging task to develop a dynamic-slope massflow algorithm for a simulation model based on an analogous principle. The use of a single approach renders the simulation program more coherent and its results more easy to understand in terms of the effects of the controlling factors. The only geometric element employed in the present model is the slope angle used to determine the starting point of massflow deposition in the 2-D space of the basinal cross-section PSEPA (see section 5.4). Otherwise, the whole massflow sedimentation is based on the diffusion equations, as described in section 5.4.1.

The choice of approach in the present case was dictated by the need to make the massflow algorithm fully compatible with the simulation master program. However, several other numerical choices had to be made, keeping in mind that a program, to be useful, should be relatively simple, while being also reasonably realistic. For example, the program would undoubtedly be simpler if the stability algorithm were left out, but this would render the whole program too simplistic and much less reliable. Another choice to make was to decide as to how many of the algorithm's parameters a modeller should be able to control through the input file. In order to make the model as flexible as possible, the author has decided that the angle of internal friction, the coefficient of consolidation, the PSEPA (for each sediment type) and the diffusion coefficients for massflow sedimentation (section 5.3.1 and 5.4) are all to be specified in the input file. In other words, the modeller has the possibility to analyze the controlling role of each of them individually or in various combinations.

7.2 Massflow Sedimentation in Sequence Stratigraphic Framework

Computer simulation models are of particular importance in sequence stratigraphic studies, and much of the sequence stratigraphic theory is, in fact, based on computer simulations (cf. Posamentier & Vail (1988)). However, the concepts of sequence stratigraphy have so far given relatively little account of massflow sedimentation, because no suitable simulation program was avialable.

The principal subaqueous environments of massflow sedimentation are deep-sea fans, as well as submarine ramps and slope aprons (Reading & Richards (1994); see also section 3.2). In other environments, massflow deposits are generally insignificant as basin-fill successions on a sequence stratigraphic scale.

The sequence stratigraphic literature has thus far considered mainly the basin-floor massflow sedimentation associated with forced regressions and lowstand wedges (Vail *et al.*, 1977; Mitchum Jr., 1984; Wagoner *et al.*, 1988; Haq, 1991; Posamentier *et al.*, 1991; Hunt & Tucker, 1992; Bowen *et al.*, 1994), which means lowstand systems tracts in the terminology of Posamentier & Vail (1988). The reason of this limited focus is more conceptual than real. When the relative sea level in a shelf/slope setting is lowered near or below the shelf break, huge amounts of sediment are eroded and delivered to the shelf edge or the adjacent slope. This may result in conditions of slope steepening and sediment overpressuring due to the high accumulation rate, leading to significant slope instability (Posamentier *et al.*, 1991). Retrogressive slope failures lead to the formation of submarine canyons, and the sediment delivered to the canyon mouth is further deposited as turbidites and/or debrites on the basin floor. The lowering of the sea level can further reduce the external hydrostatic load and create excess pore pressure in the deposits, resulting in pore-water escape and slope failures (Ouchi *et al.*, 1995). Therefore, submarine massflow sedimentation has commonly been associated with the relative fall and/or lowstand of the sea level, which means lowstand fans. Arguably, this conceptual reasoning is an oversimplification. Massflow sedimentation can occur at any time, independently of the relative sea-level stand. In a transgressive systems tract, the submarine canyon can be extended landwards by the scouring action of marine currents combined with retrogressive slumping (Steffens, 1986; Galloway *et al.*, 1991; Weimer, 1990; Helland-Hansen & Gjelberg, 1994). Massflow sedimentation in a transgressive systems tract can occur also when there is an oversupply of sediment, with the rate of sediment yield higher than the rate of accomodation space development at the shoreline. In such conditions, an active sediment transfer to the deeper-water basin can occur on a large scale (Helland-Hansen & Gjelberg, 1994). Massflow sedimentation in transgressive systems tracts has been documented by Kolla & Perlmutter (1993), Sakai & Masuda (1996) and Weber *et al.* (1997). Likewise, significant slope failures and massflow sedimentation can occur in a highstand systems tract when the basin is oversupplied with sediment and the latter is readily transferred to the deeper-water zone (Miall, 1986; Helland-Hansen & Gjelberg, 1994). Massflow sedimentation in highstand systems tracts has been documented by Sakai & Masuda (1996) and Weber *et al.* (1997).

The conclusion is that massflow sedimentation is not inherent in any particular systems tract and can occur in a wide range of basinal conditions. This is consistent with the view of Morton (1993), who has suggested that slope failures and massflow processes can occur during any phase of a relative sea-level change and, therefore, have little genetic significance within the conceptual framework of sequence stratigraphy (Morton, 1993, p. 1079). Furthermore, one has to keep in mind that the relative sea level is not the only factor controlling submarine massflow sedimentation. Computer simulations are thus of great value in allowing to assess under what conditions a particular systems tract may be prone to massflow sedimentation.

The simulation program presented in this book employs an engineering method of slope stability analysis as the chief criterion for massflow triggering. Further modelling with the use of DEMOSTRAT should reveal as to whether the conditions of slope instability are preferentially related to any specific developmental stages of particular systems tracts. An extensive simulation with different combinations of input parameters will be required for this purpose, but the results may be rewarding. It would be of crucial importance to sequence stratigraphy if the massflow sedimentation potential of various types of evolving systems tract could be assessed for various input conditions.

7.3   Limitations and Possible Adjustments

The simulation program obviously has many limitations, and some additional conceptual guidelines for the interpretation of the massflow sedimentation modelled by DEMOSTRAT may thus be useful to the reader. The deposit of a single time-step in a particular column of the basinal cross-section is regarded as a homogenous sediment package. In reality, smaller-scale heterogeneities can occur and be important in causing sediment instability. In other words, several failure planes can in reality be developed within any single time-step deposit. However, the program is a relatively large-scale model, where the time-scale limitation is not critical. The slope-stabillizing process of mass wastage of and the volumetric quantity of sediment transferred to the basin are expected to be roughly the same, no matter if there are many small massflows or a few large ones involved.

In the program, the velocity of a massflow depends upon the sediment mass involved and its boundary and internal friction characteristics (see equations 3.2 and 3.4). In general, the greater the mass, the higher the flow velocity, which means that larger massflows can be expected to have longer runout distances. The bottom friction is another controlling factor, which depends upon the seafloor roughnesses and includes topographic undulations, in the form of "throughs" and "mounds", related to the preceding depositional phases. If the cross-sectional width of a through is greater than the corresponding width of a massflow, the latter can be confined by the through and the effective bottom friction is assumed to be lower in such conditions, compared to that experienced by an unconfined massflow. The development of turbulence is another factor controlling a massflow's runout distance. Turbidites tend to be deposited more distally than debrites in submarine-fan systems. For a massflow to become turbulent, the basinal slope must be sufficiently long to allow acceleration. If the slope's gradient abruptly decreases, in the form of a "slope break", the resulting hydraulic jump may cause flow-energy dissipation (see Garcia & Parker, 1989) and instigate rapid deposition. All the controlling factors mentioned above can be accounted for, indirectly, in a simulation by adjusting the CSG or, more generally, the specification of the failure, bypass and deposition zones (used in the massflow algorithm) in the input file. However, this requires from the modeller some good fundamental knowledge on both massflow processes and the simulated basinal conditions.

The CSG can be expected to depend also upon the 3-D conditions of the simulated basinal setting. For example, the massflow deposits on submarine aprons are generally characterized by lesser basinward extent than those on submarine fans, due to the differences in the feeder system's behaviour. Furthermore, the slope-failure criterion (i.e., the critical FS-value) should be altered if the modeller expects a high-energy shoreline with large wave fetch to be involved, because wave-induced pore pressures are not directly accounted for by the instability algorithm. Again, these adjustments may require some experience or good a priori knowledge from the modeller.

The spatial pattern of massflow deposition, as shown graphically by the simulation printout, should also be considered in terms of the possible implications - which may be right or false. The massflow sedimentation in the program is governed by diffusion equations (section 5.4.1), whereby the deposit of each DEP is invariably fining upwards and basinwards. In reality, the deposit of a single DEP may consist of several DEVs and be much more complex as a whole. In other words, the deposit of a single DEV is likely to be fining upwards and basinwards, but the stacking of the deposits of successive DEVs may result in a more complex textural trend in the bulk DEP deposit. The PSEPA concept implies that the better sorted the source sediment and the smaller the massflow volumes, the more offset the successive DEV deposits will be with respect to one another, which means an increasing departure of the simulation result from the reality. In either case, massflow deposits in reality are often much less graded than the simulation program may imply. This cautionary remark pertains to all dynamic-slope simulation programs, whose aim is to model large-scale depositional episodes.

In summary, it is always advantageous for the modeller to have a good physical knowledge of the basinal setting and the massflow processes that are to be simulated, as this may help the controlling parameters to be tuned up to reality and prevent the simulation results from being misinterpreted.

7.4   Dimension Problems

In a straightforward application, a two-dimensional computer program that models the depositional system's longitudinal cross-section, with the sediment input from the margin, seems most suitable for submarine fans, which means systems related to a point source of sediment supply. The simulation results for such settings may, indeed, be fairly close to the reality, although the modeller must keep in mind the program's simplifying assumption: all sediment given as the system's input will be distributed within the 2-D space of the basinal cross-section. In a basinal reality, some of the sediment will almost invariably "escape" from the cross-section, due to the lateral shifting of the system's depositional axis or the feeding point. This means that the total amount of sediment fed to the basinal cross-section should not be assumed to represent the bulk sediment supply to the basin-margin delta or canyon. The sediment input should rather be understood as the amount of sediment deposited in a longitudinal cross-section's plane of the basin. Since this notion pertains to any depositional settings, the use of a computer program such as DEMOSTRAT is not necessarily restricted to point-source systems.

In broad terms, a stratigraphic simulation program can be thought of as a mass-budget program, with the massflow algorithm playing the role of a mass-budget balancing device. This means that the sediment that cannot be deposited stably in the canyon or shelf-margin delta has to be transferred to the deep-water basin floor, such that the sediment mass of the two parts of the system remains equal to the input mass. The mass redistribution as such can thus be modelled for virtually any basinal settings or supply conditions. Even if the simulation result fails to give an accurate picture on what one would expect in a basin's cross-section, the result is always instructive - provided that the modeller keeps in mind that the 2-D longitudinal section is actually a mass-balancing projection of the system's 3-D space. The input parameters in the program can be regarded as the strike-averages values for the feeder system, which is not an unrealistic assumptions on the account of the large time-spans used. The outlets of rivers and submarine canyons are known to shift laterally with time. For example, the Mississippi fan has been fed consecutively by no less than 17 different point sources (Weimer, 1990) during the last 3.5 mln yrs, with a lateral migration of the river mouth over a distance of 250 km. The effect of the lateral shifts is that the fan's longitudinal sections may be roughlly the same over large lateral distances. Given a mean sedimentation rate and a mean initial basin form, the simulation result will reflect the mean deposition cross-section of the system, no matter how large its strike-parallel extent may be. Taking into account the relatively long time-spans involved, the result may thus be quite realistic - if properly interpreted. As pointed out by Reading & Richards (1994), the consequences of changes in depositional systems are infinitely complex and not easily predictable. Computer simulation programs such as DEMOSTRAT may help to understand this complexity.

7.5   Future Improvements

The equation for sediment excess pore pressure (eq. 2.14) should in the future be modified to include both a varying sedimentation rate and two sediment types. It is obvious that changes in the sedimentation rate may be critical for the development of excess pore pressures. More importantly, the interpolation between the excess pore pressures in sand (assumed to be zero) and mud, based on the actual sand fraction (see Δu in section 5.3.1), may not necessarily be correct, even though it seems to be the best arbitrary approximation for the time being.

It might be desirable to have the two-dimensional model extended to three dimensions, as this would solve many of the problems discussed above (section 7.4). However, one must bear in mind that a model never is better than the assumptions it is based upon. A good and transparent 2-D model may still be better than a poor and opaque 3-D one. A significant improvement would be to allow out-of-plane deposition (i.e., sediment "escape" from the simulated basinal section). This is numerically not a difficult task, but will certainly increase the complexity of the program and render it more difficult to use.

More simulations and comparative actualistic studies are needed to evaluate the diffusion coefficients, which play a key role in the program. These coefficients specify as to how easily the erosion and deposition occur, and determine the sedimentary geometry of the basinal cross-section modelled. In the shallower-water zone of the cross-section (basin margin area), massflows are effectively smoothing out and stabilizing the failing sedimentary slope. Therefore, any diffusion-based stratigraphic program, even if not modelling massflow sedimentation explicitly, requires the shallow-zone diffusion coefficients to account for the reduction of slope. A program modelling massflow sedimentation requires the shallow-water diffusion coefficients to separate the slope-reducing effect of mass failure, because the latter is to be accounted for by the massflow processes. Consequently, the diffusion coefficients for the basin-margin zone may be different depending upon whether a program is modelling massflow sedimentation explicitly, or merely implicitly.

Some generalization of the diffusion coefficients for massflow sedimentation can be made, although the coefficient for mud should be kept larger than that for sand. This is because mud in a turbidity current tends to remain in suspension longer than sand. Likewise, mudflows commonly have longer runouts than sandflows (section 3.2). Furthermore, the coefficients may be adjusted to reflect better the feeder system. For example, the massflow deposits of submarine fan have a greater basinward extent than those of submarine ramps, and the latter are still more extensive than the massflow deposits of slope aprons. As summarized in section 3.2, the basinward lengths of massflow deposits in a submarine fan can be assumed to be from 1 km to more than 2000 km, with gradients usually about 2°-5° for the upper fan, to less than 1° for the lower fan. The lengths of massflow deposits in submarine ramps can be assumed to be between 1 and 200 km, with gradients varying from 14 to 0.14°, whereas the lengths of deposits in slope aprons can be between 1 and 100 km, with gradients varying from 26.6 to 1.15°. The diffusion coefficients have to be selected to assure that these general characteristics are met by the model, although more work is needed to understand better these coefficients in particular basinal settings.

An obvious and simple way of improving and calibrating the input parameters would through an extensive simulation scheme with the use of the program. It is meaningful that almost every computer program is sooner or later replaced with a new, improved version. Simply, the best way of improving a program is to test it extensively for a wide range of cases. In short, the present version of the simulation program is by no means final, and is certainly open to various improvements.

Chapter 8

Conclusions

In the early days of lithostratigraphic analysis, the focus of basinal studies was on relatively broad lithological categories and the available palaeontological data. Sedimentary sucessions were analyzed in terms of lithosomes (presently referred to "rock bodies") and the basin-fill architecture was studied chiefly on the basis of biostratigraphic correlations, which were the only way to recognise possible "time lines (or surfaces)" in a basin-fill succession. The modern methodology of lithofacies analysis has shifted the focus to the *dynamics* of depositional systems, a term that encompasses the process of sediment erosion, transport and deposition, but goes well beyond them, including a wider range of "authogenic" and "allogenic" controls on sedimentation. Facies analysis has led to a tremendous progress in our understanding of sedimentation, but it also made us realize that the nature of the sedimentation process is so incredibly complex, and the laboratory possibilities to simulate the multivariable "chaotic" phenomena so very limited, that the geologists have begun to be content with facies models and basin-fill facies architecture. Simultaneously, the modern methods of seismic stratigraphy have helped us to realize the importance of the information contained in basin-fill architecture. This has led to the advent of *sequence stratigraphy*, which means deciphering basin-fill architecture in terms of time surfaces that derive directly from facies analysis, rather than palaeontological data (although the latter are by no means neglected). Sequence stratigraphy relies strongly on stratigraphic *modelling*, as the means of insight in basin-fill history and

117

verification of concepts.

The aim of stratigraphic modelling is to reproduce general, qualitative trends, spatial facies organization, rather than the precise details of actual processes and short-term changes. Such apparent "softening" of the sharp focus of sedimentology does not signal, however, a diminishing ambition to understand sedimentation and basin-fill dynamics. On the contrary, it simply reflects a return to a broader basinal view and our awareness that the basin-filling depositional systems evolve under a complex influence of factors that are only partly understood. What the modelling algorithms lack in specificity and process detail, they make up for in generality. Modern sequence stratigraphy seeks to find relations among the most diverse sedimentation phenomena and their controls, by means of both case studies and simulated analogies. At that higher level of sedimentary basin analysis, lithostratigraphy has returned to the original goal of trying to understand the origin and spatial organization of lithosomes.

Computer simulations have recently become an important tool of sequence stratigraphy and basin analysis, aimed at the modelling of basin-fill sedimentary successions and their depositional architectures (e.g., Lukyanov, 1987; Helland-Hansen *et al.*, 1988; Anderson & Humphrey, 1990; Lawrence *et al.*, 1990; Syvitski & Daughney, 1992; Rivenæs, 1992, 1993). In the present volume, the stratigraphic computer-simulation program DEMOSTRAT (Rivenæs, 1993) has successfully been extended to model subaqueous siliciclastic massflow sedimentation. Submarine massflow sedimentation, although important and very common in nature, was not satisfactorily accounted for by the existing numerical programs.

The program extension is based on the equation defining the factor of safety (FS) in infinite-slope theory. The equation is solved numerically by using, among other parametres, the excess pore-pressure term of Gibson's (1958) sediment consolidation theory. The equation in the program plays the role of a principal criterion that controls the triggering of massflow processes. In the numerical framework of DEMOSTRAT, the erosion, transport and deposition of sediment by massflow processes are described by the physical equations of difusion. The present algorithm is original and quite noval in its approach. No comparable, advanced algorithm for massflow sedimentation was available in the existing programs for dynamic-slope stratigraphic modelling. The computer model pertains to longer-term increments of basinal sedimentation by massflow processes, rather than individual massflows, but this is a desirable approach and by no means a disadvantage in the present case.

The seven example simulations serve to show, in terms of graphical computer printouts, how the massflow sedimentation model operates. The results demonstrate that the model is satisfactory, although some further adjustments are recommended as a possible improvement. An extensive modelling of well-understood basin-fill sucecessions is recommended to tune up, so to speak, the parameters employed in the model. An empirically-verifiable modelling would also be the best way to assess the need for possible improvements and to implement such.

Appendix A

List of Symbols

The following symbols have been used with consistency:

Symbol		Explanation	Unit [SI unit]
A	=	Thickness of surface zone	Length [m]
C	=	Concentrations of grains (i.e. 1- porosity)	Dimensionless
D	=	Flexural rigidity of lithosphere	Force * Length [Nm]
E	=	Young's modulus	Force/Length^2 [Nm^{-2}]
F	=	Force	Force [N]
G	=	Weight	Force [N]
H	=	Wave-height	Length [m]
K	=	Transport coefficient	Area/Time [m^{2}s^{-1}]
L	=	Wave-length	Length [m]
M	=	Unidimensional compressibility	Length^2/Mass [m^{2}kg^{-1}]
N	=	Number	Dimensionless
P	=	Pressure	Force/Area [Nm^{-2}]
R	=	Channel radius	Length [m]
S	=	Shear strength	Force/Area [Nm^{-2}]
T	=	Effective lithospheric plate thickness	Length [m]
U	=	Velocity of uniform flow	Length/Time [ms^{-1}]
V	=	Volume	Volume [m^3]
W	=	Water depth	Length [m]
X	=	Vertical distance from bottom of column to where Δu is computed	Length [m]
Z	=	Total sediment thickness	Length [m]
FS	=	Factor of safety	Dimensionless

Appendix A

Symbol		Explanation	Unit [SI unit]
CSG	=	Critical sedimentation gradient	
DEP	=	Depositional episode	
DEV	=	Depositional event	
EGR	=	Episode of gradual removal	
EIR	=	Episode of instantaneous removal	
IEP	=	Instability episode	
MEV	=	Mass-gravity flow event	
PSEPA	=	Point of significant episodic aggradation	
a	=	Acceleration coefficient (expressed in terms of g)	Length/Time^2 [ms^{-2}]
c	=	Cohesion	Force/Area [Nm^{-2}]
d	=	Thickness	Length [m]
e	=	Void ratio	
f	=	Friction coefficient	Dimensionless
g	=	Gravity acceleration	Length/Time^2 [ms^{-2}]
h	=	Topographic height	Length [m]
k	=	Yield strength	Force/Area [Nm^{-2}]
l	=	Lithology flux fraction in transport relation	Dimensionless
m	=	mass	Mass [kg]
p	=	Poisson's ratio	Dimensionless
q	=	Sedimentation rate	Volume/Time [m^{3}s^{-1}]
r	=	Radius of rigid plug of debris flow	Length [m]
s	=	Sand bulk-volume fraction	Dimensionless
t	=	Time	Time [s]
u	=	Pore pressure	Force/Area [Nm^{-2}]
v	=	Velocity	Length/Time [ms^{-1}]
x	=	Spatial length ($\perp$ on y)	Length [m]
y	=	Spatial length ($\perp$ on x)	Length [m]
z	=	Sedimentary thickness	Length [m]
$zmax$	=	Maximum distance from sediment surface that FS is calculated	Length [m]
Δu	=	Pore pressure in excess of hydrostatic pressure	Force/Area [Nm^{-2}]

Symbol		Explanation	Unit [SI unit]
α	=	Slope angle	Degrees [°]
γ	=	Unit (volumetric) weight	Force/Volume [Nm^{-3}]
η	=	Eddy viscosity	Force*Time/Length^2 [Nsm^{-2}]
θ	=	Angle of repose	Degrees [°]
κ	=	Permeability	[Darcy]
μ	=	Fluid viscosity	Force*Time/Length^2 [Nsm^{-2}]
ξ	=	x/Z	Dimensionless
ρ	=	Density	Mass/Volume [kgm^{-3}]
ϱ	=	Density of pore fluid	Mass/Volume [kgm^{-3}]
σ	=	Normal stress	Force/Area [Nm^{-2}]
ς	=	Coefficient of consolidation	Area/Time [m^2s^{-1}]
τ	=	Shear stress	Force/Area [Nm^{-2}]
ϕ	=	Angle of internal friction	Degrees [°]
		Superscripts	
B	=	Constant	
$\prime$	=	With respect to effective stress	
		Subscripts	
B	=	Body of turbidity current	
H	=	Head of turbidity current	
L	=	Mud/shale indicator	
N	=	On force to indicate normal force	
S	=	Sand/sandstone indicator	
W	=	Water indicator	
b	=	Used on unit weight to indicate buoyant unit weight	
c	=	critical (on angle of repose)	
i	=	Internal	
n	=	Top boundary of flow	
0	=	Bottom boundary of flow	
r	=	Used of force to indicate resisting force	
s	=	Used of force to indicate shearing force	
w	=	Used on shear stress caused by waves	
x	=	Direction indicator	
y	=	Direction indicator	
ap	=	Apparent	
max	=	Maximum value of sth.	
min	=	Minimum value of sth.	

Appendix B

Numerical Analysis

The numerical procedure *gauher* that is used to solve the equation for the excess pore pressure (eq. 2.14) in the sediment, is a procedure that solves integrals of the form

$$\int_{-\infty}^{\infty} e^{-x^2} f(x) dx$$

To be able to use this method on equation 2.14, it must be shown that the equation is even (i.e. $f(-x) = f(x)$), since, for even functions:

$$\int_0^{\infty} f(x) dx = \frac{1}{2} \int_{-\infty}^{\infty} f(x) dx$$

In addition, it must be shown that the integrand of the equation is continuous, defined for $x = 0$ and that the integral converges. Also, a substitution must be found so that the exponential part of the equation can be expressed of the form e^{-x^2}.

Symmetry: Note first that

$$\tanh(x) = \frac{e^x - e^{-x}}{e^x + e^{-x}}$$

$$\Downarrow$$

$$\tanh(-x) = \frac{e^{-x} - e^x}{e^{-x} + e^x} = \frac{-(e^x - e^{-x})}{e^x + e^{-x}} = -\tanh(x)$$

and that

$$\cosh(x) = \frac{e^x + e^{-x}}{2}$$

$$\Downarrow$$

$$\cosh(-x) = \frac{e^{-x} + ex}{2} = \cosh(x)$$

Now turn to equation 2.14. Let

$$I(\xi) = \underbrace{\xi \tanh\left(\frac{m\xi}{2\varsigma}\right)}_{I_1(\xi)} \underbrace{\cosh\left(\frac{x\xi}{2\varsigma t}\right)}_{I_2(\xi)} \underbrace{e^{-\frac{\xi^2}{4\varsigma t}}}_{I_3(\xi)}$$

$$I_1(-\xi) = -\xi \tanh -\frac{m\xi}{2\varsigma} = (-\xi)\left(-\tanh\frac{m\xi}{2\varsigma}\right) = \xi \tanh \frac{m\xi}{2\varsigma} = I_1(\xi)$$

$$I_2(-\xi) = \cosh -\frac{x\xi}{2\varsigma t} = \cosh \frac{x\xi}{2\varsigma t} = I_2(\xi)$$

$$I_3(-\xi) = e^{-\frac{(-\xi)^2}{4\varsigma t}} = e^{-\frac{\xi^2}{4\varsigma t}} = I_3(\xi)$$

Thus

$$I(-\xi) = I_1(-\xi)I_2(-\xi)I_3(-\xi) = I_1(\xi)I_2(\xi)I_3(\xi) = I(\xi)$$

which means that equation 2.14 is even.

Continuity:
$I(\xi)$ is defined for $\xi = 0$ and consists only of continuous functions. Therefore $I(\xi)$ is continuous $\forall \xi$.

Convergence:
Since $I(\xi)$ is an even function, it is only necessary to evaluate the function at one of the limits of the integral. $I(\xi)$ is of the same form as $I(x)$ defined below

$$I(x) = x \tanh x \cosh x e^{-x^2}$$

$$\Downarrow$$

$$I(x) = x \;\; \frac{e^x - e^{-x}}{e^x + e^{-x}} \;\; \frac{e^x + e^{-x}}{2} \;\; e^{-x^2}$$

$$\Downarrow$$

$$I(x) = \frac{x(e^{2x} - e^{-2x})e^{-x^2}}{2(e^x + e^{-x})}$$

$$\Downarrow$$

$$I(x) = \frac{x(e^x - e^{-3x})e^{-x^2 - x}}{2(1 + e^{-2x})}$$

In the numerator the e^{-x^2}-part will dominate when $x \to \infty$, meaning that the numerator approaches 0 as $x \to \infty$. The denominator approaches 2 as $x \to \infty$. As a consecquence, $\int_{-\infty}^{\infty} I(x)dx$ converges and so does $\int_{-\infty}^{\infty} I(\xi)d\xi$.

Substitution:

Let

$$\zeta = \frac{\xi}{2\sqrt{\varsigma t}}$$

Then

$$\int_{-\infty}^{\infty} \xi \tanh \frac{m\xi}{2\varsigma} \cosh \frac{x\xi}{2\varsigma t} e^{-\frac{\xi^2}{4\varsigma t}} d\xi$$

$$\Updownarrow$$

$$\int_{-\infty}^{\infty} 2\sqrt{\varsigma t}\zeta \tanh \frac{m2\sqrt{\varsigma t}\zeta}{2\varsigma} \cosh \frac{x2\sqrt{\varsigma t}\zeta}{2\varsigma t} e^{-\frac{4\varsigma t\zeta^2}{4\varsigma t}} 2\sqrt{\varsigma t}d\zeta$$

This means that the integral can be written as:

$$\frac{1}{2} \int_{-\infty}^{\infty} 4\varsigma t\zeta \tanh \left(\sqrt{\frac{t}{\varsigma}}m\zeta \right) \cosh \left(\frac{x\zeta}{\sqrt{\varsigma t}} \right) e^{-\zeta^2} d\zeta$$

which is an integral that can readily be solved by the numerical method *gauher*.

A common way to numerically solve an integral of a function is to multiply the functional values at a set of equally spaced points with certain aptly chosen weighting coefficients and to take the sum of these products over the integration interval. Thus, for the improper integral:

$$\int_{-\infty}^{\infty} f(x)\, dx \approx \sum_{j=1}^{N} f(x_j)(x_{j+1} - x_j)$$

The approximation becomes less good when the integrand is $f(x)e^{-x^2}$. *Gauher* provides freedom to choose not only the weighting coefficients, but also the location of the abcissas at which the function is to be evaluated (they will no longer be equally spaced). To put it simple, *gauher* finds a weighting function, w, such that

$$\int_{-\infty}^{\infty} f(x)e^{-x^2}\, dx \approx \sum_{j=1}^{N} w_j\, f(x_j)$$

Thus, the method can be applied by defining the function f as:

$$f(\zeta) = 4\varsigma t\zeta \tanh\left(\sqrt{\frac{t}{\varsigma}}m\zeta\right) \cosh\left(\frac{x\zeta}{\sqrt{\varsigma t}}\right)$$

Bibliography

ADAMS, J. I. (1965) The engineering behaviour of a Canadian muskeg. *Proceedings of the 6th Iternational Conference on Soil Mechanics and Foundation Engineering*, **1**, 3–7.

AIGNER, T., BRANDENBURG, A., VLIET, A. V., DOYLE, M., LAWRENCE, D., & WESTRICH, J. (1990) Stratigraphic modeling of epicontinental basins. *Sedimentary Geology*, **69**, 167–190.

AKIYAMA, J. & STEFAN, H. G. (1986) Prediction of turbidity currents in reservoirs and coastal regions. In: *River Sedimentation – Volume III* (Ed. by S. Y. Wang *et al.*), pp. 1295–1305. School of Engineering, University of Mississippi.

ALLEN, J. R. L. (1971) Mixing at turbidity current heads, and its geological implications. *Journal of Sedimentary Petrology*, **41**, 97–113.

ALLEN, J. R. L. (1978) Studies in fluviatile sedimentation: An exploratory quantitative model for the architecture of avulsion-controlled alluvial suites. *Sedimentary Geology*, **21**, 129–147.

ALLEN, J. R. L., ed. (1984) *Sedimentary Structures - Their Characteristics and Physical Basis*. Elsevier. 1256 pp.

ALLEN, J. R. L., ed. (1985) *Principles of Physical Sedimentology*. George Allen and Unwin. 272 pp.

ALMAGOR, G. & WISEMAN, G. (1978) Analysis of submarine slumping in the continental slope off the southern coast of Israel. *Marine Geotechnology*, **2**, 349–388.

129

ANDERSON, R. S. & HUMPHREY, N. F. (1990) Interaction of weathering and transport processes in the evolution of arid landscapes. In: *Quantitatve Dynamic Stratigraphy* (Ed. by T. A. Cross), pp. 349–361. Prentice Hall.

ARCHER, J. S. & WALL, C. G., eds. (1986) *Petroleum Engineering Principles and Practice.* Graham & Trotman. 362 pp.

ATHANASIOU-GRIVAS, D. (1978) Reliability analysis of earth slopes. In: *Proceedings, 15th Annual Meeting*, pp. 453–458. Society of Engineering Sciences.

AUDET, D. M. & FOWLER, A. C. (1992) A mathematical model for compaction in sedimentary basins. *Geophysical Journal International*, **110**, 577–590.

BAGNOLD, R. A. (1954) Experiments on a gravity-free dispersion of large solid spheres in a Newtonian fluid under shear. *Proceedings of the Royal Society of London, Series A*, **225**, 49–63.

BAGNOLD, R. A. (1962) Auto-suspension of transported sediment: turbidity currents. *Proceedings of the Royal Society of London, Series A*, **265**, 315–319.

BAGNOLD, R. A. (1966) An approach to the sediment transport problem from general physics. *U.S. Geological Survey Professional Paper*, **422**, 37 pp.

BALDWIN, B. & BUTLER, C. O. (1985) Compaction curves. *The American Association of Petroleum Geologists Bulletin*, **69**, 622–626.

BARKER, C. (1972) Aquathermal pressuring - the role of temperature in the development of abnormal pressure zones. *The American Association of Petroleum Geologists Bulletin*, **56**, 2068–2071.

BARKER, C. & HORSFIELD, B. (1982) Mechanical versus thermal cause of abnormally high pore pressures in shales. *The American Association of Petroleum Geologists Bulletin*, **66**, 99–100.

BEEN, K. & SILLS, G. C. (1981) Self-weight consolidation of soft soils: an experimental and theoretical study. *Géotechinique*, **31**, 519–535.

BICE, D. (1988) Synthetic stratigraphy of carbonate platform and basin systems. *Geology*, **16**, 703–706.

BISHOP, A. W. (1973) The stability of tips and spoil heaps. *Quarternary Journal of Engineering Geology*, **6**, 335–376.

BITZER, K. & HARBAUGH, J. W. (1987) DEPOSIM: A Macintosh computer model for two-dimensional simulation of transport, deposition, erosion, and compaction of clastic sediments. *Computers & Geosciences*, **13**, 611–637.

BOGGS, S. J., ed. (1987) *Principles of Sedimentology and Stratigraphy*. Merrill Publishing Company. 784 pp.

BONHAM-CARTER, G. & HARBAUGH, J. W. (1968) Simulation of geologic systems: An overview. In: *Computer Contribution No. 22*, pp. 3–10. Kansas Geological Survey.

BONHAM-CARTER, G. & HARBAUGH, J. W. (1971) Stratigraphic modeling by compter simulation. In: *Data Processing in Biology and Geology* (Ed. by J. L. Cutbill), pp. 147–164. Academic Press.

BOOTH, J. S. (1979) Recent history of mass-wasting on the upper continental slope, Northern Gulf of Mexico, as interpreted from the consolidation states of the sediment. *Society of Economic Paleontologists and Mineralogists Special Publication*, **27**, 153–164.

BOOTH, J. S., SANGREY, D. A., & FUGATE, J. K. (1985) A nomogram for interpreting slope stability of fine-grained deposits in modern and ancient marine environments. *Journal of Sedimentary Petrology*, **5**, 29–36.

BOUMA, A. H., DeVRIES, M. B., & STONE, C. G. (1997) Reinterpretation of depositional processes in a classic flysch sequence (Pennsylvanian Jackfork Group), Ouachita Mountains, Arkansas and Oklahoma: discussion. *The American Association of Petroleum Geologists Bulletin*, **81**, 470–472.

BOUMA, A. H., NORMARK, W. R., & BARNES, N. E., eds. (1985) *Turbidite Fans and Related Turbidite Systems*. Springer-Verlag. 351 pp.

BOWEN, D. W., WEIMER, P., & SCOTT, A. J. (1994) The relative success of silisiclastic sequence stratigraphic consepts in exploration: Examples from incised valley fill and turbidite systems reservoirs. In: *Silisiclastic Sequence Stratigraphy: Recent Developments and Applications* (Ed. by P. Weimer & H. W. Posamentier), pp. 15–42. American Association of Petroleum Geologists Memoir, **58**.

BRIDGE, J. S. & LEEDER, M. R. (1979) A simulation model of alluvial stratigraphy. *Sedimentology*, **26**, 617–644.

BRUNSDEN, D. & PRIOR, D. B., eds. (1984) *Slope Instability*. John Wiley & Sons Ltd. 620 pp.

BUCHAN, S., DEWES, F. C. D., MCCANN, D. M., & SMITH, D. T. (1967) Measurements of the acoustic and geotechnical properties of marine sediment cores. In: *Marine Géotechinique* (Ed. by A. F. Richards), pp. 65–92. University of Illinois Press.

BURTON, R., KENDALL, C. G. S. C., & LERCHE, I. (1987) Out of our depth: On the impossibility of fathoming eustacy from the stratigraphic record. *Earth-Science Reviews*, **24**, 237–277.

BUSCH, W. H. & KELLER, G. H. (1983) Analysis of sediment stability on the Peru-Chile continental slope. *Marine Geotechnology*, **5**, 181–211.

CANT, D. J. (1989) Simple equations of sedimentation: Applications to sequence stratigraphy. *Basin Research*, **2**, 73–81.

CANT, D. J. (1991) Geometric modelling of facies migration: Theoretical development of facies successions and local unconformities. *Basin Research*, **3**, 51–62.

CAO, S. & LERCHE, I. (1994) A quantitative model of dynamic sediment deposition and erosion in three dimensions. *Computers & Geosciences*, **20**, 635–663.

CARRIGY, M. A. (1970) Experiments on the angles of repose of granular materials. *Sedimentology*, **14**, 147–158.

CARSON, M. A. & KIRKBY, M. J., eds. (1972) *Hillslope Form and Process*. Cambridge University Press. 475 pp.

CASAGRANDE, A. (1936) Characteristics of cohesionless soils affecting the stability of slopes and earth fills. In: *Boston Society of Civil Engineers, Contributions to Soil Mechanics, 1925 - 1940*, pp. 257 – 276. Spaulding-Moss.

CHIKITA, K. (1990) Sedimentation by river-induced turbidity currents: field measurements and interpretation. *Sedimentology*, **37**, 891–905.

CHOUGH, S. & HESSE, R. (1976) Submarine meandering talweg and turbidity currents flowing for 4000 km in the Northwest Atlantic Mid-ocean Channel, Labrador Sea. *Geology*, **4**, 529–533.

CHRISTIE-BLICK, N. (1991) Onlap, offlap, and the origin of unconformity-bounded depositional sequences. *Marine Geology*, **97**, 35–56.

COAKLEY, B. J. & WATTS, A. B. (1991) Tectonic controls on the development of unconformities: The North Slope, Alaska. *Tectonics*, **10**, 101–130.

COLELLA, A. & PRIOR, D. B., eds. (1990) *Coarse-Grained Deltas*, vol. 10. International Association of Sedimentologists Special Publication. 357 pp.

COLEMAN JR., J. L. (1997) Reinterpretation of depositional processes in a classic flysch sequence (Pennsylvanian Jackfork Group), Ouachita Mountains, Arkansas and Oklahoma: discussion. *The American Association of Petroleum Geologists Bulletin*, **81**, 466–469.

COLLIER, R. E. L., LEEDER, M. R., & MAYNARD, J. R. (1990) Transgressions and regressions: A model for the influence of tectonic subsidence, deposition and eustasy, with application to

Quarternary and Carboniferous examples. *Geological Magazine*, **127**, 117–128.

COLLINS, K. & MCGOWN, A. (1974) The form and function of microfabric features in a variety of natural soils. *Géotechinique*, **24**, 223–254.

COLLINSON, J. D. & THOMPSON, D. B., eds. (1989) *Sedimentary Structures*. Chapman & Hall. 207 pp.

COLTEN-BRADLEY, V. A. (1987) Role of pressure in smectite dehydration - effects on geopressure and smectite-illite transformation. *The American Association of Petroleum Geologists Bulletin*, **71**, 1414–1427.

COULOMB, C. A. (1776) Essais sur une application des règles des maximis et minimis à quelques problems de statique relatifs à l'architecture. Mémoirs Academie des Sciences, Paris.

CRAIG, R. F., ed. (1987) *Soil Mechanics*. Van Nostrand Reinhold, 4th edn. 410 pp.

CURRAY, J. R. & MOORE, D. G. (1971) Growth of the Bengal Deep-Sea Fan and denundation in the Himalayas. *Geological Society of America Memoir*, **84**, 563–572.

DADE, W. B., LISTER, J. R., & HUPPERT, H. E. (1994) Fine-sediment deposition from gravity surges on uniform slopes. *Journal of Sedimentary Research*, **A64**, 423–432.

D'AGOSTINO, A. E. & JORDAN, D. W. (1997) Reinterpretation of depositional processes in a classic flysch sequence (Pennsylvanian Jackfork Group), Ouachita Mountains, Arkansas and Oklahoma: discussion. *The American Association of Petroleum Geologists Bulletin*, **81**, 473–475.

DAINES, S. (1982) Aquathermal pressuring and geopressure evaluation. *The American Association of Petroleum Geologists Bulletin*, **66**, 931–939.

DONOVAN, D. T. & JONES, E. J. W. (1979) Causes of world-wide changes in sea-level. *Journal of the Geological Society of London*, **136**, 187–192.

DUBRULE, O. (1988) A review of stochastic models for petroleum reservoirs. In: *Proc. International Meeting on Quantification of Sediment Body Geometries and their Internal Heterogeneities*. British Society of Reservoir Geologists.

EDGERS, L. (1981) Viscous analysis of submarine flows. Tech. Rep. 52207-3, Norwegian Geotechnical Institute. 39 pp.

EDGERS, L. & KARLSRUD, K. (1982) Soil flows generated by submarine slides - case studies and consequenses. In: *International Conference on the Behaviour of Offshore Structures*, pp. 425–437. BOSS '82 Proceedings, **3**.

EDGERS, L. & KARLSRUD, K. (1986) Viscous analysis of submarine flows. *Norwegian Geotecnical Bulletin*, **143**, 1–10.

FERTL, W. H. (1973) Significance of shale gas as an indicator of abnormal pressures. *Society of Petroleum Engineers Paper*, **4230**.

FLEMINGS, P. B. & JORDAN, T. E. (1989) A synthetic stratigraphic model of foreland basin development. *Journal of Geophysical Research*, **94**, 3851–3866.

FROHLICH, C. & MATTHEWS, R. K. (1991) STRATA-VARIOUS: A flexible Fortran program for dynamic forward modeling of stratigraphy. In: *Sedimentary Modeling: Computer Simulations and Methods for Improved Parameter Definition* (Ed. by E. K. Franseen, W. L. Watney, C. G. S. C. Kendall, & W. Ross), pp. 449–461. Kansas Geological Survey, Bulletin, **233**.

FROSSARD, A. (1979) Effect of sand grain shape on interparticle friction; indirect measurements by Rowe's stress dilatancy theory. *Géotechinique*, **29**, 341–350.

GAFFIN, S. R. & MAASCH, K. A. (1991) Anomalous cyclicity in climate and stratigraphy and modeling nonlinear oscillations. *Journal of Geophysical Research*, **96**, 6701–6711.

Bibliography

GALLOWAY, W. E., DINGUS, W. F., & PAIGE, R. E. (1991) Seismic and depositional facies of Paleocene - Eocene Wilcox Group submarine canyon fills, northwest Gulf Coast, USA. In: *Seismic Facies and Sedimentary Processes of Submarine Fans and Turbidite Systems* (Ed. by P. Weimer & M. H. Link), pp. 247 – 271. Springer Verlag.

GARCIA, M. & PARKER, G. (1989) Experiments on hydraulic jumps in turbidity currents near a canyon-fan transition. *Science*, **245**, 393–396.

GIBSON, R. E. (1958) The progress of consolidation in a clay layer increasing in thickness with time. *Géotechinique*, **8**, 171–182.

GILLOTT, J. E., ed. (1968) *Clay in Engineering Geology*. Elsevier Publishing Company. 296 pp.

GORSLINE, D. S. & EMERY, K. O. (1959) Turbidity-current deposits in San Pedro and Santa Monica basins off southern California. *Geological Society of America Bulletin*, **70**, 279–290.

GOSH, J. K., MAZUMDER, B. S., SAHA, M. R., & SENGUPTA, S. (1986) Deposition of sand by suspension currents: experimental and theoretical studies. *Journal of Sedimentary Petrology*, **56**, 57–66.

GRAY, D. H. & LEISER, A. T., eds. (1982) *Biotechnical Slope Protection and Erosion Control*. Van Nostrand Reinhold Company. 271 pp.

GRETENER, P. E. (1981) Pore pressure: fundamentals, general ramifications, and implications for structural geology (revised). *Ameerican Association of Petroleum Geologists Education Cource Notes Series*, **4**. 131 pp.

HALDORSEN, H. H. & LAKE, L. W. (1990) Stochastic modeling. *Journal of Petroleum Technology*, **42**, 404–412.

HAMPTON, M. A. (1972) The role of subaqueous debris flow in generating turbidity currents. *Journal of Sedimentary Petrology*, **42**, 775–793.

HAMPTON, M. A., BOUMA, A. H., CARLSON, P. R., MOLINA, B. F., CLUKEY, E. C., & SANGREY, D. A. (1978) Quantitative study of slope instability in The Gulf of Alaska. *Proceedings of the 10th Offshore Technology Conference*, **4**, 2307–2318.

HAMPTON, M. A., LEE, H. J., & LOCAT, J. (1996) Submarine landslides. *Reviews of Geophysics*, **34**, 33–59.

HANSHAW, B. & ZEN, E. (1965) Osmotic equilibrium and overthrust faulting. *The American Association of Petroleum Geologists Bulletin*, **76**, 1379–1387.

HAQ, B. U. (1991) Sequence stratigraphy, sea-level change, and significance for the deep sea. *In: International Association of Sedimentologists Special Publication*, **12**, 3–39.

HARBAUGH, J. W. (1966) Mathematical simulation of marine sedimentation with IBM 7090/7094 computers. *Computer Contribution No. 1*, pp. 1–52. Kansas Geological Survey.

HARBAUGH, J. W. & BONHAM-CARTER, G., eds. (1970) *Computer Simulation in Geology*. Wiley Interscience. 575 pp.

HARBAUGH, J. W. & BONHAM-CARTER, G. (1977) Computer simulation of continental margin sedimentation. In: *The Sea, Volume 6: Marine Modeling* (Ed. by Goldberg *et al.*), pp. 623–649. John Wiley & Sons.

HARDIN, B. O. (1989) Low-stress dilation test. *Journal of Geotechnical Engineering*, **115**, 769–787.

HEDBERG, H. D. (1974) Relation of methane generation to undercompacted shales, shale diapirs and mud volcanoes. *The American Association of Petroleum Geologists Bulletin*, **58**, 661–673.

HEEZEN, B. C. & HOLLISTER, C. D., eds. (1971) *The Face of the Deep*. Oxford University Press. 659 pp.

HELLAND-HANSEN, W. & GJELBERG, J. G. (1994) Conceptual basis and variability in sequence stratigraphy: A different perspective. *Sedimentary Geology*, **92**, 31–52.

HELLAND-HANSEN, W., KENDALL, C. G. S. C., LERCHE, I., & NAKAYAMA, K. (1988) A simulation of continental basin margin sedimentation in responce to crustal movements, eustatic sea level change, and sediment accumulation rates. *Mathematical Geology*, **20**, 777–802.

HELLAND-HANSEN, W. & MARTINSEN, O. J. (1997) Shorline trajectories and sequences: Description of variable depositional-dip scenarios. *Journal of Sedimentary Research*, **66**, 670–688.

HELLER, P. L. & DICKINSON, W. R. (1985) Submarine ramp facies model for delta-fed, sand-rich turbidite systems. *The American Association of Petroleum Geologists Bulletin*, **69**, 960–976.

HENKEL, D. J. (1970) The role of waves in causing submarine landslides. *Géotechinique*, **20**, 75–80.

HISCOTT, R. N., PICKERING, K. T., BOUMA, A. H., HAND, B. M., KNELLER, B. C., POSTMA, G., & SOH, W. (1997) Basin-foor fans in the North Sea: sequence stratigraphic models vs. sedimentary facies: discussion. *The American Association of Petroleum Geologists Bulletin*, **81**, 662–665.

HORN, H. M. & DEERE, D. U. (1962) Frictional characteristics of minerals. *Géotechinique*, **12**, 319–335.

HOROWITZ, D. H. (1976) Mathematical modeling of sediment accumulation in prograding deltaic systems. In: *Quantitative Techniques for the Analysis of Sediments* (Ed. by D. F. Merriam), pp. 105–119. Pergamon Press.

HOWELL, D. G. & NORMARK, W. R. (1982) Sedimentology of submarine fans. *The American Association of Petroleum Geologists Memoir*, **31**, 365–404.

HUNT, D. & TUCKER, M. E. (1992) Stranded parasequences and the forced regressive wedge systems tract: Deposition during base level fall. *Sedimentary Geology*, **81**, 1–9.

INOUCHI, Y., KINUGASA, Y., KUMON, F., NAKANO, S., YASUMATSU, S., & SHIKI, T. (1996) Turbidites as records of intense palaeoearthquakes in Lake Biwa, Japan. *Sedimentary Geology*, **104**, 117–125.

JERVEY, M. T. (1988) Quantitative geological modeling of silisi-clastic rock sequenses and their seismic expression. In: *Sea-Level Changes: An Integrated Approach* (Ed. by C. K. Wilgus, B. S. Hastings, C. G. S. C. Kendall, H. W. Posamentier, C. A. Ross, & J. C. V. Wagoner), pp. 47–69. Society of Economic Paleontologists and Mineralogists Special Publication 42.

JIBSON, R. W. (1992) The Mameyes, Puerto Rico, landslide disaster of October 7, 1985. In: *Landslides/Landslide Mitigation* (Ed. by J. E. Slosson, A. G. Keene, & J. A. Johnson), pp. 37–54. Geological Society of America, Reviews in Engineering Geology, **9**.

JOHNSON, A. M., ed. (1970) *Physical Processes in Geology*. Freeman Publication Company. 571 pp.

JOHNSON, A. M. (1984) Debris flows. In: *Slope Instability* (Ed. by D. Brunsden & D. B. Prior). John Wiley & Sons.

JOHNSON, A. M. & RODINE, J. R. (1984) Debris flows. In: *Slope Instability* (Ed. by D. Brunsden & D. B. Prior), pp. 257–361. John Wiley & Sons.

JONES, M. (1994) Mechanical principles of sediment deformation. In: *The Geological Deformation of Sediments* (Ed. by A. Maltman), pp. 37–71. Chapman & Hall.

JORDAN, T. E. & FLEMINGS, P. B. (1991) Large-scale stratigraphic architecture, eustatic variation, and unsteady tectonism: A theoretical evaluation. *Journal of Geophysical Research*, **96**, 6681–6699.

KARNER, G. D. & DRISCOLL, N. W. (1997) Three-dimensional interplay of advective and diffusive processes in the generation of sequence boundaries. *Journal of the Geological Society*, **154**, 443–449.

Bibliography

KAUFMAN, P., GROTZINGER, J. P., & McMORMICK, D. S.
(1991) Depth-dependent diffusion algorithm for simulation of sedimentation in shallow marine depositional systems. In: *Sedimentary Modeling: Computer Simulations and Methods for Improved Parameter Definition* (Ed. by E. K. Franseen, W. L. Watney, C. K. C. G., St., & W. Ross), pp. 489–508. Kansas Geological Survey, Bulletin, **233**.

KELTS, K. & ARTHUR, M. A. (1981) Turbidites after ten years of deep-sea drilling - wringing out the mop? In: *The Deep Sea Drilling Project: A Decade of Progress* (Ed. by J. E. Warme, R. G. Douglas, & E. L. Winterer), pp. 91–127. Society of Economic Paleontologists and Mineralogists Special Publication **32**.

KENDALL, C. G. S. C., STROBEL, J., CANNON, R., BEZDEK, J., & BISWAS, G. (1991) The simulation of the sedimentary fill of basins. *Journal of Geophysical Research*, **96**, 6911–6929.

KENNEY, C. (1984) Properties and behaviours of soils relative to slope instability. In: *Slope Instability* (Ed. by D. Brunsden & D. B. Prior), pp. 27–66. John Wiley & Sons.

KOLLA, V. & PERLMUTTER, M. A. (1993) Timing of turbidite sedimentation on the Mississippi Fan. *The American Association of Petroleum Geologists Bulletin*, **77**, 1129–1141.

KOLMOGOROV, A. N. (1951) Solution of a problem in probability theory connected with the problem of mechanism of stratification. *Transactions of the American Mathematical Society*, **53**, 171–177.

KOMAR, P. D. (1969) The channelized flow of turbidity currents with application to Monterey deep-sea fan channel. *Journal of Geophysical Research*, **74**, 4544–4558.

KONDNER, R. L. & VENDRELL JR., J. R. (1964) Consolidation coefficients: Cohesive soil mixtures. *Journal of Soil Mechanics and Foundation Division, A. S. C. E.*, **90**, 31–42.

KOSTASCHUK, A. & McCANN, B. (1989) Submarine slope stability of a fjord delta: Bella Coola, British Columbia. *Geographie physique et Quaternaire*, **43**, 87–95.

140

KUENEN, P. H. (1967) Emplacement of flysch-type sand beds. *Sedimentology*, **9**, 203–243.

LADE, P. V. (1993) Initiation of static instability in the submarine Nerlerk berm. *Canadian Geotechnical Journal*, **30**, 895–904.

LAMBE, T. W. & WHITMAN, R. V., eds. (1979) *Soil Mechanics.* Massachusetts Institute of Technology, John Wiley & Sons Inc. 553 pp.

LAWRENCE, D. T., DOYLE, M., & AIGNER, T. (1990) Stratigraphic simulation of sedimentary basins. *The American Association of Petroleum Geologists Bulletin*, **74**, 273–295.

LEE, H. J., CHOUGH, S. K., CHUN, S. S., & HAN, S. J. (1991) Sediment failure on the Korea Plateau slope, East Sea (Sea of Japan). *Marine Geology*, **97**, 363–377.

LEE, H. J., CHOUGH, S. K., & YOON, S. H. (1996) Slope-stability change from late Pleistocene to Holocene in the Ulleung Basin, East Sea (Japan Sea). *Sedimentary Geology*, **104**, 39–51.

LEE, H. J., CHUN, S. S., YOON, S. H., & KIM, S. R. (1993) Slope stability and geotechnical properties of sediment of the southern margin of Ulleung Basin, East Sea (Sea of Japan). *Marine Geology*, **110**, 31–45.

LEE, H. J. & EDWARDS, B. D. (1986) Regional method to assess offshore slope stability. *Journal of Geotechnical Engineering*, **112**, 489–509.

LEONARDS, G. A. & RAMIAH, B. K. (1960) Time effect in the consolidation of soils. In: *Papers on Soils 1959 Meetings*, pp. 116–130. American Society for Testing and Materials, Special Technical Publication, **254**.

LERCHE, I. (1990) Philosophies and strategies of model building. In: *Quantitative Dynamic Stratigraphy* (Ed. by T. A. Cross), pp. 21–44. Prentice Hall.

LO, S.-C., CHEN, M.-P., & FAN, J.-C. (1997) Slope stability and geotechnical properties of sediment off the Changyuan area, Eastern Taiwan. *Marine Georesources and Geotechnology*, **15**, 209–229.

LØNNE, I. (1995) Sedimentary facies and depositional architecture of ice-contact glacimarine systems. *Sedimentary Geology*, **98**, 13–43.

LØNNE, I. (1997) Facies characteristics of a proglacial turbiditic sand-lobe at Svalbard. *Sedimentary Geology*, **109**, 13–35.

LOWE, D. R. (1976) Subaquous liquified and fluidized sediment flows and their deposits. *Sedimentology*, **23**, 285–308.

LOWE, D. R. (1982) Sediment gravity flows: II. Depositional models with special reference to the deposits of high-density turbidity currents. *Journal of Sedimentary Petrology*, **52**, 279–297.

LOWE, D. R. (1997) Reinterpretation of depositional processes in a classic flysch sequence (Pennsylvanian Jackfork Group), Ouachita Mountains, Arkansas and Oklahoma: discussion. *The American Association of Petroleum Geologists Bulletin*, **81**, 460–465.

LUKYANOV, A. V. (1987) Self-exited oscillations in geological systems (model studies and problems of correlation). In: *Global Correlation of Tectonic Movements* (Ed. by Y. G. Leonov & V. E. Khain), pp. 231–272. John Wiley & Sons Ltd.

LUTHI, S. (1980) Some new aspects of two-dimensional turbidity currents. *Sedimentology*, **28**, 97–105.

LUTHI, S. (1981) Experiments on non-channelized turbidity currents and their deposits. *Marine Geology*, **40**, M59–M68.

MAGARA, K. (1975) Importance of the aquathermal pressuring effect in the Gulf Coast. *The American Association of Petroleum Geologists Bulletin*, **59**, 2037–2045.

MALTMAN, A., ed. (1994) *The Geological Deformation of Sediments*. Chapman & Hall. 362 pp.

MANDL, G. & CRANS, W. (1981) Gravitational gliding in deltas. In: *Trust and Nappe Tectonics* (Ed. by K. R. McClay & N. J. Price), pp. 41–55. Geological Society of London Special Publication 9.

MARTINSEN, O. (1994) Mass movements. In: *The Geological Deformation of Sediments* (Ed. by A. Maltman), pp. 127–165. Chapman & Hall.

MERIFIELD, P. M. (1992) Surficial slope failures in southern California Willside residental areas: Lessons from the 1978 and 1980 rainstorms. In: *Landslides/Landslide Mitigation* (Ed. by J. E. Slosson, A. G. Keene, & J. A. Johnson), pp. 37–54. Geological Society of America, Reviews in Engineering Geology, **9**.

MIALL, A. D. (1986) Eustatic sea level changes interpreted from seismic stratigraphy: A critique of the methodology with particular reference to the North Sea Jurassic record. *The American Association of Petroleum Geologists Bulletin*, **70**, 131–137.

MIDDLETON, G. V. (1966a) Experiments on density and turbidity currents: I. Motion of the head. *Canadian Journal of Earth Sciences*, **3**, 523–546.

MIDDLETON, G. V. (1966b) Experiments on density and turbidity currents: II. Uniform flow of density currents. *Canadian Journal of Earth Sciences*, **3**, 627–637.

MIDDLETON, G. V. (1969) Grain flows and other mass movements down slopes. In: *The New Concepts of Continental Margin Deposition* (Ed. by D. J. Stanley), pp. GM-B-1 to GM-B-14. American Geological Institute, Short Course Lecture Notes.

MIDDLETON, G. V. (1970) Experimental studies related to flysch sedimentation. In: *Flysch Sedimentology in North America* (Ed. by J. Lajoie), pp. 253–272. The Geological Association of Canada, Special Paper, **7**.

MIDDLETON, G. V. (1976) Hydraulic interpretation of sand size distributions. *Journal of Geology*, **84**, 405–426.

Bibliography

MIDDLETON, G. V. & HAMPTON, M. (1976) Subaqueous sediment transport and deposition by sediment gravity flows. In: *Marine Sediment Transport and Environmental Management* (Ed. by D. J. Stanley & D. J. P. Swift), pp. 197–218. Wiley (Interscience).

MIDDLETON, G. V. & NEAL, W. J. (1989) Experiments on the thickness of beds deposited by turbidity currents. *Journal of Sedimentary Petrology*, **59**, 297–307.

MIDDLETON, G. V. & SOUTHARD, J. B. (1984) Mechanics of sediment movement. *Society of Economic Paleontologists and Mineralogists, Short Course Lecture Note*, **3**. Tulsa.

MIDDLETON, G. V. & WILCOCK, P. R., eds. (1994) *Mechanics in the Earth an Environmental Sciences*. Cambridge University Press.

MILLER, R. L. & BYRNE, R. J. (1966) The angle of repose for a single grain on a fixed rough bed. *Sedimentology*, **6**, 303–314.

MITCHELL, J. K., ed. (1993) *Fundamentals of Soil Behaviour*. John Wiley & Sons, 2nd edn.

MITCHUM JR., R. M. (1984) Seismic stratigraphic criteria for recognition of submarine fans. *Society of Economic Paleontologists and Mineralogists, Gulf Coast Section, Annual Research Conference, Program and Abstracts*, pp. 63–85.

MORGENSTERN, N. R. (1967) Submarine slumping and the initiation of turbidity currents. In: *Marine Géotechinique* (Ed. by A. F. Richards), pp. 189–220. University of Illinois Press.

MORTON, R. A. (1993) Attributes and origins of ancient submarine slides and filled embayments: Examples from the Gulf Coast Basin. *The American Association of Petroleum Geologists Bulletin*, **77**, 1064–1081.

MULDER, T., BERRY, J. A., & PIPER, D. J. W. (1997a) Links between morphology and geotechnical characteristics of large debris flow deposits in the Albatross area on the Scotian Slope (SE Canada). *Marine Georesources and Geotechnology*, **15**, 253–281.

MULDER, T., SAVOYE, B., & SYVITSKI, J. P. M. (1997b) Numerical modelling of a mid-sized gravity flow: the 1979 Nice turbidity current (dynamics, processes, sediment budget and seafloor impact). *Sedimentology*, **44**, 305–326.

MURRAY, T. (1994) Glacial deformation. In: *The Geological Deformation of Sediments* (Ed. by A. Maltman), pp. 73–93. Chapman & Hall.

MUTTI, E. (1985) Turbidite systems and their relations to depositional sequences. In: *Provenance of Arenites* (Ed. by G. G. Zuffa), pp. 65–93. Reidel.

MUTTI, E. & NORMARK, W. R. (1987) Comparing examples of modern and ancient turbidite systems: Problems and concepts. In: *Marine Clastic Sedimentology: Concepts and Case Studies* (Ed. by J. K. Leggett & G. G. Zuffa), pp. 1–38. Graham & Trotman.

MUTTI, E. & RICCI-LUCCHI, R. (1972) Turbidites of the northern Apennines: Introduction to facies analysis. *International Geological Reviews*, **20**, 125–166.

MUTTI, E. & RICCI-LUCCHI, R. (1975) Turbidite facies and facies associations. *In: Field Trip Guidebook, 9th. Int. Sedimentology Congress*, **A-11**, 21–36.

NACCI, V. A., KELLY, W. E., WANG, W. C., & DEMARS, K. R. (1974) Strength and stress-strain characteristics of cemented deep-sea sediments. In: *Deep-Sea Sediments: Physical and Mechanical Properties* (Ed. by A. L. Inderbitzen), pp. 129–150. Plenum Press.

NELSON, C. H. & KULM, L. D. (1973) Submarine fans and channels. In: *Society of Economic Paleontologists and Mineralogists Short Course Lecture Notes*, pp. 39–78. Anaheim.

NELSON, C. H. & NILSEN, T. H. (1974) Depositional trends of modern and ancient deep-sea fans. In: *Modern and Ancient Geosynclinal Sedimentation* (Ed. by R. H. Dott Jr. & R. H. Shaver), vol. 19, pp. 69–91. Society of Economic Paleontologists and Mineralogists Special Publication.

NELSON, R. B. & LINDSLEY-GRIFFIN, N. (1987) Biopressured carbonate turbidite sediments: a mechanism for submarine slumping. *Geology*, **15**, 817–820.

NEMEC, W. (1990) Aspects of sediment movement on steep delta slopes. In: *Coarse-Grained Deltas* (Ed. by A. Colella & D. B. Prior), pp. 29–73. The International Association of Sedimentologists, Special Publication, **10**.

NEMEC, W. & POSTMA, G. (1991) Inverse grading in gravel beds. In: *Abstract I. A. S. 12th. Regional Meeting*, p. 38. University of Bergen.

NEMEC, W. & STEEL, R. J. (1984) Alluvial and coastal conglomerates: Their significant features and some comments on gravelly mass-flow deposits. In: *Sedimentology of Gravels and Conglomerates* (Ed. by E. H. Koster & R. J. Steel), pp. 1–31. Memoir of the Canadian Society of petroleum Geologists ,**10**.

NILSEN, T. (1980) Modern and ancient submarine fans: Discussion of papers by R. G. Walker and W. R. Normark. *The American Association of Petroleum Geologists Bulletin*, **64**, 1094–1112.

NITZSCHE, M. (1989) Submarine slope instability, Eastern Banda Sea. *Netherlands Journal of Sea Research*, **24**, 431–436.

NORMARK, W. R. (1978) Fan valleys, channels, and depositional lobes on modern submarine fans: Characters for recognition of sandy turbidite environments. *The American Association of Petroleum Geologists Bulletin*, **62**, 912–931.

NORMARK, W. R. (1989) Observed parameters for turbidity-current flow in channels, Reserve Fan, Lake Superior. *Journal of Sedimentary Petrology*, **59**, 423–431.

NORMARK, W. R., POSAMENTIER, H., & MUTTI, E. (1993) Turbidite systems: state of the art and future directions. *Review of Geophysics*, **31**, 91–116.

146

OLSON, R. E. (1974) Shearing strengths of kaolinite, illite and montmorillonite. *Journal of the Geotechnical Engineering Division, American Socitey of Civil Engineers*, **100**, 1215–1229.

OUCHI, S., ETHRIDGE, F. G., JAMES, E. W., & SCHUMM, S. A. (1995) Experimental study of subaqueous fan development. In: *Characterization of Deep Marine Clastic Systems* (Ed. by A. J. Hartley & D. J. Prosser), pp. 13–29. Geological Society Special Publication, **94**.

PANTIN, H. M. (1979) Interaction between velocity and effective density in turbidity flow: phase-plane analysis, with criteria for autosuspension. *Marine Geology*, **31**, 55–99.

PAOLA, C., HELLER, P. L., & ANGEVINE, C. A. (1992) The large-scale dynamics of grain-size variation in alluvial basins, 1: Theory. *Basin Research*, **4**, 73–90.

PAPATHEODOROU, G. & FERENTINOS, G. (1997) Submarine and coastal sediment failure triggered by the 1995, $M_S = 6.1$ Aegion earthquake, Gulf of Corinth, Greece. *Marine Geotechnology*, **137**, 287–304.

PARKER, G. (1982) Conditions for the ignition of catastrophically erosive turbidity currents. *Marine Geology*, **46**, 307–327.

PIPER, D. J. W., SHOR, A. N., & HUGHES CLARKE, J. E. (1988) The 1929 Grand Banks earthquake, slump and turbidity current. *Geological Society of America Bulletin*, **229**, 77–92.

PIPER, D. J. W., STOW, D. A. V., & NORMARK, W. R. (1984) The Laurentian Fan: Sohm Abyssal Plain. *Geo-Marine Letters*, **3**, 141–146.

PITMAN, W. C. & GOLOVCHENKO, X. (1983) The effect of sea level change on the shelfedge and slope of passive margins. In: *The Shelfbreak: Critical Interface on Continental Margins* (Ed. by D. J. Stanley & G. T. Moore), pp. 41–58. Society of Economic Paleontologists and Mineralogists Special Publication, **3**.

PLINT, A. G., EYLES, N., EYLES, C. H., & WALKER, R. G. (1992) Controls of sea level change. In: *Facies Models, Responce to Sea Level Change* (Ed. by R. G. Walker & N. P. James), pp. 15–25. Geological Association of Canada.

POSAMENTIER, H. W., ERSKINE, R. D., & MITCHUM JR., R. M. (1991) Models for submarine fan deposition within a sequence-stratigraphic framework. In: *Seismic Facies and Sedimentary Processes of Submarine Fans and Turbidite Systems* (Ed. by P. Weimer & H. Link), pp. 127–136. Springer-Verlag.

POSAMENTIER, H. W. & VAIL, P. R. (1988) Eustatic controls on clastic deposition, II. Sequence and systems tract models. In: *Sea-Level Changes - An Integrated Approach* (Ed. by C. K. Wilgus, B. S. Hastings, C. A. Ross, H. W. Posamentier, J. C. V. Wagoner, & C. G. C. S. Kendall), pp. 125–154. Society of Economic Paleontologists and Mineralogists, Special Publication, **42**.

POULOS, S. J., CASTRO, G., & FRANCE, J. W. (1985) Liquefaction evaluation procedure. *Journal of Geotechnical Engineering*, **111**, 772–791.

PRESS, W. H., TEUKOLSKY, S. A., VETTERLING, W. T., & FLANNERY, B. P., eds. (1992) *Numerical Recipies in Fortran*. Cambridge University Press, 2nd edn. 963 pp.

PRIOR, D. B. & COLEMAN, J. M. (1984) Submarine slope instability. In: *Slope Instability* (Ed. by D. Brunsden & D. B. Prior), pp. 419–455. John Wiley & Sons.

PRIOR, D. B., COLEMAN, J. M., SUHAYDA, J. N., & GARRISON, L. E. (1979) Subaqueous landslides as they affect bottom structures. In: *Port and Ocean Engineering Under Artic Conditions, at the Norwegian Institute of Technology*, pp. 921–933. Coastal Studies Institute.

READING, H. G. & RICHARDS, M. (1994) Turbidite systems in deep-water basin margins classified by grain size and feeder system. *The American Association of Petroleum Geologists Bulletin*, **78**, 792–822.

RICHARDS, A. F. & HAMILTON, E. L. (1967) Investigations of deep-sea sediment cores, III. Consolidation. In: *Marine Géotechinique*, pp. 93–117. University of Illinois Press.

RIVENÆS, J. C. (1992) Application of a dual-lithology, depth-dependent diffusion equation in stratigraphic simulation. *Basin Research*, **4**, 133–146.

RIVENÆS, J. C. (1993) *A computer simulation model for silisiclastic basin stratigraphy*. Ph.D. thesis, Norwegian Institute of Technology, University of Trondheim. 133 pp.

ROBERTS, J. A. & CRAMP, A. (1996) Sediment stability on the western flank of the Canary Islands. *Marine Geology*, **134**, 13–30.

ROSS, W. C. (1990) Modeling base-level dynamics as a control on basin-fill geometries and facies distribution: A conceptual framework. In: *Quantitatve Dynamic Stratigraphy* (Ed. by T. A. Cross), pp. 387–399. Prentice Hall.

ROSS, W. C., WATTS, D. E., & MAY, J. A. (1995) Insights from stratigraphic modeling: Mud-limited versus sand-limited depositional systems. *The American Association of Petroleum Geologists Bulletin*, **79**, 231–258.

ROTHMAN, D. H., GROTZINGER, J. P., & FLEMINGS, P. (1994) Scaling in turbidite deposition. *Journal of Sedimentary Research*, **A64**, 59–67.

RUPKE, N. A. & STANLEY, D. J., eds. (1974) *Distinctive properties of turbiditic and hemipelagic mud layers in the Algero-Balearic Basin, Western Mediterranean Sea*. Smithsonian Contribution to Earth Sciences, **13**.

SAKAI, T. & MASUDA, F. (1996) Slope turbidite packets in a fore-arc basin fill sequence of the Plio-Pleistocene Kakegawa Group, Japan: their formation and sea-level changes. *Sedimentary Geology*, **104**, 89–98.

SANGREY, D. A. & MARKS, D. L. (1981) Hindcasting evaluation of slope stability in the Baltimore Canyon Trough area. *13th Annual Offshore Technology Conference*, pp. 241–248.

SAXOV, S. & NIEUWENHUIS, J. K., eds. (1982) *Marine Slides and Other Mass Movements*. NATO Conference Series IV (Marine Sciences), Plenum Press. 353 pp.

SCHANZ, T. & VERMEER, P. A. (1996) Angles of friction and dilatancy of sand. *Géotechnique*, **46**, 145–151.

SCHOFIELD, A. & WROTH, P., eds. (1968) *Critical State Soil Mechanics*. McGraw-Hill. 310 pp.

SCHWAB, W. C., LEE, H. J., & MOLINA, B. F. (1987) Causes of varied sediment gravity flow types on the Alsek prodelta, Northeast Gulf of Alaska. *Marine Geotechnology*, **7**, 317–342.

SCOTT, A. M. & BRIDGWATER, J. (1975) Interparticle percolation: a fundamental solids mixing mechanism. *Industrial Engineering Chemistry, Fundamentals*, **14**, 22–26.

SEED, H. & RAHMAN, M. S. (1978) Wave-induced pore pressure in relation to ocean floor stability of cohesionless soils. *Marine Geotechnology*, **3**, 123–150.

SHANMUGAM, G. (1996) High-density turbidity currents: are they sandy debris flows? *Journal of Sedimentary Research*, **66**, 2–10.

SHANMUGAM, G., BLOCH, R. B., MITCHELL, M., BEAMISH, G. W. J., HODGKINSON, R. J., DAMUTH, J. E., STRAUME, T., SYVERTSEN, S. E., & SHIELDS, K. E. (1995) Basin-floor fans in the North Sea: sequence stratigraphic models vs. sedimentary facies. *The American Association of Petroleum Geologists Bulletin*, **79**, 477–512.

SHANMUGAM, G. & MOIOLA, R. J. (1988) Submarine fans: characteristics, models, classification, and reservoir potential. *Earth Science Reviews*, **24**, 383–428.

SHANMUGAM, G. & MOIOLA, R. J. (1997) Reinterpretation of depositional processes in a classic flysch sequence (Pennsylvanian Jackfork Group), Ouachita Mountains, Arkansas and Oklahoma. *The American Association of Petroleum Geologists Bulletin*, **79**, 672–695.

SHIMIZU, M. (1982) Effect of overconsolidation on dilatancy of a cohesive soil. *Soils and Foundations*, **22**, 121–135.

SINCLAIR, H. D., COACLEY, B. J., ALLEN, P. A., & WATTS, A. B. (1991) Simulation of forland basin stratigraphy using a diffusion model of mountain belt uplift and erosion: An example from the central Alps, Switzerland. *Tectonics*, **10**, 599–620.

SKEMPTON, A. W. (1964) Long-term stability of clay slopes. *Géotechinique*, **14**, 77–102.

SKEMPTON, A. W. & BISHOP, A. W. (1954) Building Materials. In: *Soils* (Ed. by M. Reiner), chap. 10. North-Holland Publishing Company.

SKEMPTON, A. W. & DELORY, F. A. (1957) Stability of natural slopes in London Clay. *In: Proceedings of 4th International Conference on Soil Mechanics and Foundation Engineering*, **2**, 378–381.

SLATT, R. M., WEIMER, P., & STONE, C. G. (1997) Reinterpretation of depositional processes in a classic flysch sequence (Pennsylvanian Jackfork Group), Ouachita Mountains, Arkansas and Oklahoma: discussion. *The American Association of Petroleum Geologists Bulletin*, **81**, 449–459.

SLINGERLAND, R., HARBAUGH, J. W., & FURLONG, K., eds. (1994) *Simulating Clastic Sedimentary Basins*. Prentice Hall. 220 pp.

SOUTHARD, J. B. & MACKINTOSH, M. E. (1981) Experimental test on autosuspension. *Earth Surface Processes and Landforms*, **6**, 103–111.

STACEY, M. W. (1982) *A theoretical study of density and turbidity currents*. Ph.D. thesis, Dalhouise University.

STACEY, M. W. & BOWEN, A. J. (1988) The vertical structure of turbidity currents and a necessary condition for self-maintenance. *Journal of Geophysical Research*, **93**, **C4**, 3543–3553.

STEFFENS, G. S. (1986) Pleistocene entrenched valley/canyon systems, Gulf of Mexico (abs.). *The American Association of Petroleum Geologists Bulletin*, **70**, 1189.

STOW, D. A. V. (1981) Laurentian Fan: morphology, sediments, processes and growth pattern. *The American Association of Petroleum Geologists Bulletin*, **65**, 375–393.

STOW, D. A. V. (1984) Cretaceous to Recent submarine fans in the SE Angolan Basin. In: *Initial Report: Deep Sea Drilling Project* (Ed. by W. W. Hay & J. C. Sibuet), vol. 75. U.S. Govt. Printing Office, Washington.

STOW, D. A. V. (1985) Deep-sea clastics: where are we and where are we going? In: *Sedimentology: Recent Developments and Applied Aspects* (Ed. by P. J. Brenchley & B. P. J. Williams), pp. 67–93. Geological Society of London Special Publication, 18.

STOW, D. A. V. (1986) Deep clastic seas. In: *Sedimentary Environments and Facies* (Ed. by H. G. Reading), pp. 399–444. Blackwell Scientific Publications.

STOW, D. A. V. (1994) Deep sea processes of sediment transport and deposition. In: *Sediment Transport and Depositional Processes* (Ed. by K. Pye), pp. 257–291. Blackwell Scientific Publications.

STOW, D. A. V. & BOWEN, A. J. (1980) A physical model for the transport and sorting of fine- grained sediments by turbidity currents. *Sedimentology*, **27**, 31–46.

STOW, D. A. V., READING, H. G., & COLLINSON, J. D. (1996) Deep seas. In: *Sedimentary Environments: Processes, Facies and*

Stratigraphy (Ed. by H. G. Reading), pp. 395–453. Blackwell Science.

STROBEL, J., CANNON, R., KENDALL, C. G. S. C., BISWAS, G., & BEZDEK, J. (1989) Interactive (SEDPAK) simulation of clastic and carbonate sediments in shelf to basin settings. *Computers & Geosciences*, **15**, 1279–1290.

SYVITSKI, J. P. M. & DAUGHNEY, S. (1992) Delta2: Delta progradation and basin filling. *Computers & Geosciences*, **18**, 839–897.

SYVITSKI, J. P. M., SMITH, J. N., CALABRESE, E. A., & BOUDREAU, B. P. (1988) Basin sedimentation and the growth of prograding deltas. *Journal of Geophysical Research*, **93**, 6895–6908.

TERZAGHI, K. (1955) Influence of geological factors on the engineering properties of sediments. *Economic Geology*, **50**, 557–618.

TERZAGHI, K. (1962) Stability of steep slopes on hard unweathered rock. *Géotechinique*, **12**, 251–270.

TETZLAFF, D. M. & HARBAUGH, J. W., eds. (1989) *Simulating Clastic Sedimentation*. Van Nostrand Reinhold. 202 pp.

THORNE, J. A. & SWIFT, D. J. P. (1991) Sedimentation on continental margins, II: Application of the regime concept. In: *Shelf Sand and Sanstone Bodies: Geometry, Facies and Sequence Stratigraphy* (Ed. by D. J. P. Swift, G. F. Oertel, R. W. Tillman, & J. A. Thorne), pp. 33–58. Special Pulication of the International Association of Sedimentologists, **14**.

TIPPER, J. C. (1992) Landforms developing and basins filling: Three dimensional simulation of erosion, sediment transport, and deposition. In: *Computer Graphics in Geology* (Ed. by R. Pflug & J. W. Harbaugh), vol. 41, pp. 155–170. Springer-Verlag.

TURCOTTE, D. L. & KENYON, P. M. (1984) Synthetic passive margin stratigraphy. *The American Association of Petroleum Geologists Bulletin*, **68**, 768–775.

TURCOTTE, D. L. & SCHUBERT, G., eds. (1982) *Geodynamics - Aplications of Continuum Physics to Geological Problems.* John Wiley & Sons. 450 pp.

TURNER, J. S., ed. (1973) *Buoyancy Effects in Fluids.* Cambridge University Press. 367 pp.

VAIL, P. R., MITCHUM JR., R. M., & THOMPSON, S. (1977) Seismic stratigraphy and global changes of sea level, Part 4: Global cycles of relative changes of sea level. In: *Seismic Stratigraphy – Applications to Hydrocarbon Exploration* (Ed. by C. E. Peyton), pp. 83–97. American Association of Petroleum Geologists Memoir 26.

VAN BURKALOW, A. (1945) Angle of repose and angle of sliding friction: An experimental study. *Bulletin of the Geological Society of America,* **56,** 669–708.

VAN DER KNAAP, W. & EIJPE, R. (1968) Some experiments on the genesis of turbidity currents. *Sedimentology,* **11,** 115–124.

WAGONER, J. C. V., POSAMENTIER, H. W., MITCHUM, R. M., VAIL, P. R., SARG, J. F., LOUTIT, T. S., & HARDENBOL, J. (1988) An overview of the fundamentals of sequence stratigraphy and key definitions. In: *Sea-Level Changes - An Integrated Approach* (Ed. by C. K. Wilgus, B. S. Hastings, C. A. Ross, H. W. Posamentier, J. C. V. Wagoner, & C. G. C. S. Kendall), pp. 109–124. Society of Economic Paleontologists and Mineralogists, Special Publication, **42.**

WALKER, R. G. (1978) Deep-water sandstone facies and ancient submarine fans: models for exploration for stratigraphic traps. *The American Association of Petroleum Geologists Bulletin,* **62,** 932–966.

WALKER, R. G. (1980) Modern and ancient submarine fans. *The American Association of Petroleum Geologists Bulletin,* **64,** 1101–1108.

WALKER, R. G. & JAMES, N. P., eds. (1992) *Facies Models: Response to Sea-level Change.* Geological Association of Canada. 409 pp.

WALTHAM, D. (1992) Mathematical modeling of sedimentary basin processes. *Marine & Petroleum Geology*, **9**, 265–273.

WALTON, O. R. (1983) Particle-dynamics calculations of shear flow. In: *Mechanics of Granular Materials: New Models and Constitutive Relations* (Ed. by J. T. Jenkins & M. Satake), pp. 327–338. Elsevier.

WATTS, A. B. (1989) Lithospheric flexure due to prograding sediment loads: Implications for the origin of offlap/onlap patterns in sedimentary basins. *Basin Research*, **2**, 133–144.

WATTS, A. B. & THORNE, J. (1984) Tectonics, global changes in sea level and their relationship to stratigraphical sequences at the US Atlantic continental margin. *Marine & Petroleum Geology*, **1**, 319–339.

WEBER, M. E., WIEDICKE, M. H., KUDRASS, H. R., HUBSCHER, C., & ERLENKEUSER, H. (1997) Active growth of the Bengal Fan during sea-level rise and highstand. *Geology*, **25**, 315–318.

WEIMER, P. (1990) Sequence stratigraphy, facies geometrics, and depositional history of the Mississippi Fan, Gulf of Mexico. *The American Association of Petroleum Geologists Bulletin*, **74**, 425–453.

WHEELER, H. E. (1958) Time stratigraphy. *The American Association of Petroleum Geologists Bulletin*, **42**, 1047–1063.

WHEELER, H. E. (1964) Base level transit cycle. In: *Symposium on Syclic Sedimentation* (Ed. by D. F. Merriam), pp. 623–630. Bulletin of the Kansas Geological Survey, **169**.

WHITMAN, R. V. (1985) On liquefaction. In: *Proceedings of the 11th International Conference on Soil Mechanics and Foundation Engineering, San Fransisco*, vol. 4, pp. 1923–1926. A. A. Balkema.

Bibliography

WHITTEN, E. H. T. (1964) Process-responce models in geology. *Geological Society of America Bulletin*, **75**, 455–464.

WU, T. H., ed. (1966) *Soil Mechanics*. Allyn & Bacon. 431 pp.

YOUNG, A. & LOW, P. (1965) Osmosis in agrillaceous rocks. *The American Association of Petroleum Geologists Bulletin*, **49**, 1005–1007.

ZENG, J. (1992) *Numerical Simulation of Turbidity Current Flow and Sedimentation*. Ph.D. thesis, Department of Geology, Stanford University. 300 pp.

ZENG, J., LOWE, D. R., PRIOR, D. B., WISEMAN JR., W. J., & BORNHOLD, B. D. (1991) Flow properties of turbidity currents in Bute Inlet, British Columbia. *Sedimentology*, **38**, 975–996.

Lecture Notes in Earth Sciences

For information about Vols. 1–19
please contact your bookseller or Springer-Verlag

Vol. 57: E. Lallier-Vergès, N.-P. Tribovillard, P. Bertrand (Eds.), Organic Matter Accumulation. VIII, 187 pages. 1995.

Vol. 58: G. Sarwar, G. M. Friedman, Post-Devonian Sediment Cover over New York State. VIII, 113 pages. 1995.

Vol. 59: A. C. Kibblewhite, C. Y. Wu, Wave Interactions As a Seismo-acoustic Source. XIX, 313 pages. 1996.

Vol. 60: A. Kleusberg, P. J. G. Teunissen (Eds.), GPS for Geodesy. VII, 407 pages. 1996.

Vol. 61: M. Breunig, Integration of Spatial Information for Geo-Information Systems. XI, 171 pages. 1996.

Vol. 62: H. V. Lyatsky, Continental-Crust Structures on the Continental Margin of Western North America. XIX, 352 pages. 1996.

Vol. 63: B. H. Jacobsen, K. Mosegaard, P. Sibani (Eds.), Inverse Methods. XVI, 341 pages, 1996.

Vol. 64: A. Armanini, M. Michiue (Eds.), Recent Developments on Debris Flows. X, 226 pages. 1997.

Vol. 65: F. Sansò, R. Rummel (Eds.), Geodetic Boundary Value Problems in View of the One Centimeter Geoid. XIX, 592 pages. 1997.

Vol. 66: H. Wilhelm, W. Zürn, H.-G. Wenzel (Eds.), Tidal Phenomena. VII, 398 pages. 1997.

Vol. 67: S. L. Webb, Silicate Melts. VIII. 74 pages. 1997.

Vol. 68: P. Stille, G. Shields, Radiogenetic Isotope Geochemistry of Sedimentary and Aquatic Systems. XI, 217 pages. 1997.

Vol. 69: S. P. Singal (Ed.), Acoustic Remote Sensing Applications. XIII, 585 pages. 1997.

Vol. 70: R. H. Charlier, C. P. De Meyer, Coastal Erosion – Response and Management. XVI, 343 pages. 1998.

Vol. 71: T. M. Will, Phase Equilibria in Metamorphic Rocks. XIV, 315 pages. 1998.

Vol. 72: J. C. Wasserman, E. V. Silva-Filho, R. Villas-Boas (Eds.), Environmental Geochemistry in the Tropics. XIV, 305 pages. 1998.

Vol. 73: Z. Martinec, Boundary-Value Problems for Gravimetric Determination of a Precise Geoid. XII, 223 pages. 1998.

Vol. 74: M. Beniston, J. L. Innes (Eds.), The Impacts of Climate Variability on Forests. XIV, 329 pages. 1998.

Vol. 75: H. Westphal, Carbonate Platform Slopes – A Record of Changing Conditions. XI, 197 pages. 1998.

Vol. 76: J. Trappe, Phanerozoic Phosphorite Depositional Systems. XII, 316 pages. 1998.

Vol. 77: C. Goltz, Fractal and Chaotic Properties of Earthquakes. XIII, 178 pages. 1998.

Vol. 78: S. Hergarten, H. J. Neugebauer (Eds.), Process Modelling and Landform Evolution. X, 305 pages. 1999.

Vol. 79: G. H. Dutton, A Hierarchical Coordinate System for Geoprocessing and Cartography. XVIII, 231 pages. 1999.

Vol. 80: S. A. Shapiro, P. Hubral, Elastic Waves in Random Media. XIV, 191 pages. 1999.

Vol. 81: Y. Song, G. Müller, Sediment-Water Interactions in Anoxic Freshwater Sediments. VI, 111 pages. 1999.

Vol. 82: T. M. Løseth, Submarine Massflow Sedimentation. IX, 156 pages. 1999.